BIOFORTEAN NOTES

BIOFORTEAN NOTES

VOLUME 1

CHAD ARMENT, EDITOR

COACHWHIP PUBLICATIONS
Landisville, Pennsylvania

BioFortean Notes, Volume 1

CoachwhipBooks.com

ISBN 1-61646-087-3
ISBN-13 978-1-61646-087-7

CONTENTS

Putting the "ology" into Cryptozoology

Charles Paxton

Centre for Research into Ecological and Environmental Modelling University of St Andrews, The Observatory, Buchanan Gardens, St Andrews, Fife KY16 9LZ, UK.

Cryptozoology is sometimes criticised by scientists and skeptics as a pseudoscience (Shermer 1997, Radford & Nickell 2006, Prothero 2007) and for having a weak evidence base (Naish 2000). Paxton (2002) suggested that cryptozoology was not currently a science but was, at its best, a valid form of scholarship, akin to natural history in that it could be a scholarly investigation of the natural world, albeit without experimentation or falsification of hypotheses. Other authors have expressed similar opinions (Arment 2004, Roesch & Moore 2002). Bille (pers. comm.) has argued that many cryptozoological hypotheses are technically falsifiable (i.e. there is a large bipedal ape living in a particular forest) albeit not practically falsifiable because of budgetary constraints.

Since Paxton's (2002) article, a number of peer-reviewed papers on cryptozoological themes have been published. Some of those papers are on what might be called predictive zoological discovery, i.e. pre-discovery, submitting non-physical evidence for the existence (or extension of range in time and space) of a particular species of animal. Often such papers do not self-identify as being cryptozoologically themed (e.g. Jaric et al. 2009, Fitzpatrick et al. 2005, Roberts & Solow 2003, Boyd & Stansfield 1998), although sometimes they do (e.g. Woodley et al 2008). There have also been peer-reviewed publications that have used molecular genetic techniques to determine the affinities of physical remains of cryptozoological interest (e.g. Coltman & Davis 2005, Pierce et al. 2004, Carr et al. 2002). Meanwhile Paxton's (2002) claim that cryptozoology was not a science was actually undermined by his own (2009) demonstration that some data from cryptozoological reports are actually amenable to rigorous statistical scrutiny, in contrast to some skeptical views of cryptozoology (e.g. Shermer 2003). Paxton (2009) suggested that if available reported cryptozoological accounts can be reasonably assumed to be an unbiased sample of the total reported cryptozoological reports, then inferences can be reached as to the cryptozoological reporting process, albeit not necessarily what was seen.

However there remains a wide gap between the approaches of these papers and those published in the non-peer-reviewed literature. A peer-reviewed paper has, unsurprisingly, to pass through peer review—i.e. a manuscript will be sent to other researchers to read and comment on. If they don't like it, it will be rejected or modifications will be requested. Thus the paper needs to be well written and well reasoned. And for scientists that means very well written and very well reasoned. No, I mean it, really well reasoned. The following tips are to supply guidance to those wanting to publish papers in the mainstream peer-reviewed scholarly literature. They also may provide food for thought for authors wanting to produce informal but nonetheless scientifically rigorous publications too. It is not a general guide but deals with the specific methodological assumptions and techniques that should be employed by authors interested in cryptozoology. It is designed to augment the many points discussed in Chad Arment's *Cryptozoology: Science and Speculation*.

1. Do not assume a cryptid exists—the scientific consensus does not accept this and will reject this as a premise. The author may have compelling personal experience of the cryptid under consideration but this cannot be

used as a premise. (It can, of course, be used as evidence). This does not make scientists bigots, just people who want to have verifiable, objective evidence and work from what is known. There is nothing to stop a paper arguing for the existence of an unknown species but there must be compelling evidence with all alternative explanations for the evidence adequately considered. Evidence is not the same as proof (in a cryptozoological context, a corpse, available for analysis by independent parties) and authors should not confuse the two. Authors should also avoid making themselves hostages to fortune by confessing a personal belief in a particular cryptid in the absence of definitive proof. Scientific papers are not really the place for such comments. At best the readers will regard such a statement as an irrelevance; at worst they will consider such a statement as indicative of, ahem, monstrous subjectivity.

2. Do not assume the identity of what was reported (especially if that presumed identity is one not accepted by the scientific community as a whole, e.g. an extant Mesozoic marine reptile or Oligocene whale). Of course, authors can tentatively reach a conclusion as to identity based on the available evidence. My co-authors and I have published papers on the sea monk described in 1546 (Paxton & Holland 2005) and the monster reported by Hans Egede from 1734 (Paxton et al. 2005) and we reached conclusions as to what might have been seen, but we cannot *know* definitively what was seen. No one ultimately will know, unless the report is exposed as a hoax or a corpse turns up. Language should not be employed that gives the impression the author somehow "knows" without equivocation what was seen or what is out there, even if that is not the intention.
3. Eyewitness testimony is inaccurate, sometimes biased, and sometimes even totally false. The evidence for this

is overwhelming and has been shown again and again in a variety of psychological experiments (e.g. see Loftus 1996 for a review). One cannot discount this. However just because witnesses are inaccurate or even inconsistent does not necessarily mean they are lying.

4. The very fact that the witnesses have reported something unknown to them means they do not know what they have seen, therefore the witnesses are not really in the position to gauge their own accuracy, no matter their confidence in their own abilities. Researchers may have to assume that witnesses are right in their zoological identification or description but we should always be aware that this may not be so. They may not have even identified organs and limbs correctly (e.g. perhaps Paxton et al. 2005, Woodley et al. in press).
5. What is reported is different to what is remembered which is different to what was seen which is different to what was present.
6. It follows from 5, that the raw data of testimony-based cryptozoology is not cryptids, it is not sightings, it is reports. Reports are, of course, a biased sample of sightings which are a biased sample of what was seen. There is evidence to support this contention (see Paxton 2009). Thus reports, alas, tell us first and foremost about the reporting process and only secondarily about "cryptids."
7. Just because reports are named collectively as coming from the same general location does not necessarily mean they have the same source.
8. Just because reports are named collectively as coming from the same general cryptid, does not necessarily mean they have the same source.
9. Closely associated with point 8, beware of the cryptozoological version of the *interpretatio graeca* or *interpretatio romani*, the Greek and Roman tendency (e.g. in Herodotus' *The Histories*) to assume that the gods of

foreign non-Indo-European pantheons were analogous to those from the classical pantheon. Likewise cryptozoologists find affinities between folkloric monsters which may not reflect a real zoological or indeed folkloric connection. Although Occam's razor (i.e. when considering hypotheses of equal explanatory power choose the simplest) is a useful guide, in the case of the hypothesis that a monster has a common source as opposed to two independent origins, we cannot necessarily assume that an animal described in folklore under one name in one language is necessarily the same entity in a different language even if they appear superficially morphologically similar. For example just because someone at some point said that the thing called by the Chinese "*lóng*" is the same as the thing described in western European languages as "dragon," "*draco*" etc., does not make it so.

3–9 can be thought of avoiding *Monster Manual* thinking (Paxton 2010, but see also Woodley 2010 for a different view), that is, the tendency of cryptozoologists to regard analysis of reports/evidence secondary to the (subjective) classification, cataloguing and analysis of putative cryptids. Closely related to this is the dubious practice of allowing preconceptions to affect the interpretations of reports and how they are classified, what Woodley (2008) termed the "Oudemans fallacy."

10. The zoological community as a whole does not use the term "cryptid." That does not mean it should not be used, just that it should be defined.
11. Most zoologists have not heard of Bernard Heuvelmans, Ivan Sanderson, Loren Coleman, or any other cryptozoological personalities. And even if they do, their opinion carries no weight, evidence does.
12. As Naish (2010, 2000) has argued, be aware of the latest findings in the scientific literature. If it is thought one

class of sea monster reports are based on sightings of plesiosaurs, then make the comparison of what we know of plesiosaurs from the current peer-reviewed palaeontological literature today, not of their status in the twentieth century from coffee-table picture books. Failure to consider existing knowledge will cause a paper to be rejected.

13. Wherever possible quantify. Systematic quantification decreases the probability of subjectivity and bias (Schick & Vaughan 2002).
14. Give the wider relevance of why the research is interesting in addition to the fact it might be evidence of an unknown species (i.e. implications for eyewitness psychology, ecology of predators, palaeontology etc.).
15. In addition to not assuming a cryptid exists (see rule 1), do not premise a paper on any assumption that is not accepted by the general scientific community without some evidence to support it. This does not mean non-consensual assumptions cannot be made but that if such an assumption is made, it should be justified by reference to evidence. For example, if an author believes that certain bipedal tracks are made by an extant prehistoric animal that palaeontologists generally assume is quadrupedal, then some evidence for the presumed animal's bipedality, *independent* of the tracks, should be given.
16. The limited evidence there is (Paxton 2009), suggests that first-hand cryptozoological reports are less exaggerated than second-hand reports.
17. The limited evidence there is (Paxton 2009), suggests that anonymous cryptozoological reports are more exaggerated than reports from a named source.
18. Be aware of sampling biases. Reports are biased in a variety of ways, some of which we can predict, some of which are unknown. If monsters are primarily reported on summer days, does this mean they are more active in the summer or that there are more people to report

them? Likewise if sea monsters are primarily seen in calm weather, does this reflect a preference for clement weather or greater detection in flat calm conditions? In fact we know from marine animal survey data that cetaceans for example are much easier to detect under good conditions (e.g. fin whales from ships: Víkingsson et al. 2009). Likewise if there are historical trends in numbers of sea monster reports, one hypothesis that has to be considered is whether there are more/less people to report them. We also know in the case of marine monster reports, the data appears to be biased in that reported distances are closer than might be expected by chance alone (Paxton 2009).

19. By extension from 19, raw quantification of reports *cannot* be used to infer trends or geographic distributions or habitat preferences in the absence of some measure of search effort except in very, very, crude terms (e.g. if reports only came from mountains where few people live then it would be reasonable, *if* the reports can be assumed to come from a common source animal, to assume the animal genuinely occurred only in the mountains). Just as in any scientific animal survey, what is reported is a function not only of what is there to be seen but the effort used to hunt for it. Typically we don't know the search effort (e.g. the number of man-hours available to observe cryptids). Anecdotal reports can be considered to be a special class of what is referred by some ecologists as "presence-only" data.* Inferring habitat preferences from presence-only data can cause substantial bias (Paxton & Donovan in preparation).

*A bias caused by using the equivalent of presence-only data, was strongly implicated in the decision to launch the space shuttle *Challenger* under unfavourable environmental conditions leading ultimately to its destruction (Dawes 2001).

This point, alas, is more than occasionally missed in mainstream ecology papers written by people who really should know better.

Neglect of search effort can be seen as a special case of what is sometimes referred to in the critical thinking literature as "ratio biases," the tendency of humans to evaluate ratios/fractions inappropriately (e.g. Reyna & Brainerd 2008), although in this case the problem is a failure to recognise that a fraction should be used at all.

If search effort is measured, not reporting a cryptid can be just as informative as reporting a cryptid, at least with regard to indicating potential habitat preferences. Fortunately, with the increasing availability of motion-activated wildlife cameras, encounter per unit effort data should be available to terrestrial cryptozoologists in remote locations.

It should be noted that whilst encounter/report etc. per unit effort is better than quantifying raw reports/encounters if evaluating habitat preferences etc., even per unit effort data may have problems. It is well known from the fisheries literature where catch per unit effort (CPUE) is a commonly used statistic, that sometimes even CPUE does relate consistently to abundance (e.g. Jennings et al. 2001).

It is not just the effort of cryptozoologists directly that needs to be considered, reporting effort, that is the effort used in soliciting and collecting reports, similarly needs to be quantified. Cryptid flaps could be a function of increased cryptid activity/numbers, or more witnesses arriving to try to see the monster, but above all, of course, witnesses being more likely to consider reporting a strange encounter and the effort/interest of reporters in tackling the stories.

20. Detectability of animals is a major undiscussed issue in the search for unknown species but there is a large statistical literature on this topic (e.g. Buckland et al.

2001). Whether a species is known or unknown will be a function of the individual animal detectability (assuming people are in the right habitat), the size of the population, and the search effort in the right habitat.

21. There is evidence that both temporal and spatial trends in anomalous event reports may reflect culture rather than what was actually seen (e.g. *ghosts*: Davies 2007, Finucane 1996; *Bigfoot*: Regal 2010; *aquatic monsters*: Meurger & Gagnon 1988, Magin 1996). Thus a change in frequency of reports of one type (in the unlikely event "types" can readily be objectively identified, see Paxton 2010 and above), may not reflect a change in biology. Cryptozoologists should not ignore this.
22. The conclusions must derive from the evidence presented in the paper and the reader should be able to easily follow the argument from data/evidence to conclusion.
23. Follow the journal conventions in writing the paper.
24. Every point made *must* be justified either through evidence supplied within the paper or by reference to existing published literature. Assertion in the absence of evidence is not acceptable in peer-reviewed papers. References used should be scholarly where possible and definitely from print media or an online equivalent. Ephemeral references (websites etc.) should not be used, save as an absolute last resort. Blogs and the like are generally unacceptable as sources in formal papers. *Wikipedia* is completely unacceptable; it can be written by 12-year-olds. Getting good quality references is hard work but it is an essential part of the compilation of a scientific paper.
25. A good scientist is always honest about potential weaknesses/gaps in his assumptions, data, reasoning, or conclusions. Better the author exposes problems than the editors or referees!
26. Professional scientists may be more happy to collaborate (e.g. Crother et al. 2009) than cryptozoologists

might expect, but only if no unjustified conclusions are being made. Therefore it is worth thinking about contacting a professional scientific co-author who can help in the research and writing up. Indeed, professional zoologists may use the expertise of amateurs to contribute to their endeavours (e.g. Paxton et al. 2005). I would hope that the presence of an author from a recognised "institution" on the author list would not alter the probability of acceptance of the paper one jot, but, alas, this may not be the case, thus giving a further reason for such collaboration. One catch is that many scientists have to justify their time and so any cryptozoological work may have to be undertaken in their free time and they may not have access to resources for expensive laboratory work in what is not their core role.

27. Think hard about the appropriate journal for the material under consideration. The description of a *bona fide* Bigfoot, with the corpse deposited in a publicly accessible museum (an essential step), really should go to *Nature* or *Science*, the pre-eminent science journals in the world. Most papers will not have that impact! Interpretation of folklore/ethnoknowledge should go to a folklore or perhaps anthropology journal. Interpretation of historical accounts should go perhaps to a history of science journal. Statistical analysis of eyewitness accounts could go to general zoology journals or, depending on emphasis, psychological journals. Identification of corpses/analyses of biological material by molecular genetics means can be published in genetics journals. There is also the *Journal of Scientific Exploration*, which will take cryptozoological material, as well as the dedicated cryptozoology journal *Kraken*.
28. A paper may well be rejected. This may be because it is fundamentally flawed, badly written, or the authors were just unlucky with the review process and the reviewers. If you don't think it is fundamentally flawed,

then consider submitting it to another journal. Do not engage in a battle with the editor—that is ungentlemanly, and is guaranteed not to be productive. If you do submit to another journal, do take into account the reviewers' remarks. I have had papers rejected, sometimes unfairly in my view, but the reviewers' remarks have always proved in whole or part useful in improving the manuscript. Only once (in six submissions of material to journals) have I encountered behaviour that might be suggestive of an unfair prejudice against cryptozoological data. Otherwise editors have been happy to consider my manuscripts.

29. Remember, cryptozoology is a controversial subject, which means that cryptozoological manuscripts will receive more than the usual critical scrutiny. This possibly means that a cryptozoological paper really will need to be better than the average mainstream paper to get published.

One final seemingly self-evident point should be made. Papers must be submitted in order to be considered for publication! Cryptozoologists cannot complain about scientists ignoring their work if they don't actually submit said work for scientific consideration. So start writing!

ACKNOWLEDGEMENTS

Many thanks to Michael Woodley, Matt Bille, and Chad Arment for comments on the manuscript, especially points 26 and 27 which were Matt's and Chad's ideas!

REFERENCES

Arment, C. (2004) *Cryptozoology, Science and Speculation.* Coachwhip: Landisville.

Boyd, I. J., and Stanfield, M. P. (1998) Circumstantial evidence for the presence of monk seals in the West Indies. *Oryx* **32**, 310 – 316.

Buckland, S. T., Anderson, D. R., Burnham, K. P., Laake, J. L., Borchers, D. L., and Thomas, L. (2001) *An Introduction to Distance Sampling*. Oxford: Oxford.

Carr, S. M., Marshall, H. D., Johnstone K. A., Pynn L. M., and Stenson, G. B. (2002) How to tell a sea monster: molecular discrimination of large marine animals of the north Atlantic. *Biological Bulletin* **202,** 1 – 5.

Coltman, D., and Davis, C. (2005) Molecular cryptozoology meets the Sasquatch. *Trends in Ecology and Evolution* **21,** 60 – 61.

Crowther, B. I., White, M. E., Gardner, D., and Warms, J. (2009) Giant Canadian snakes and forensic phylogenetics. *Contemporary Herpetology* **2002(2),** 1 – 4.

Davies, O. (2007) *The Haunted. A Social History of Ghosts.* Palgrave Macmillan: London.

Dawes, R. M. (2001) *Everyday Irrationality*. Westview. Oxford.

Finucane, R. C. (1996) *Ghosts: Appearances of the Dead & Cultural Transformation.* Prometheus: New York.

Fitzpatrick, J. W., Lammertink, M., Luneau, M. D., Gallagher, T. W., Harrison, B.R., Sparling, G. M., Rosenberg, K. V., Rohrbaugh, R. W., Swarthout, E. C. H., Wrege, P. H., Swarthout, S. B., Dantzker, M. S., Charif, R. A., Barksdale, T. R., Remsen, J. V., Simon, S. D., and Zollner, D. (2005) Ivory-billed woodpecker (*Campephilus principalis*) persists in continental North America. *Science* **308,** 1460 – 1462.

Jaric, I., Lenhardt, M., Cvijanovic, G., and Ebenhard, T. (2009). *Acipenser sturio* and *Acipenser nudiventris* in the Danube – extant or extinct? *Journal of Applied Ichthyology.* **25,** 137 – 141.

Jennings, S., Kaiser, M. J., and Reynolds, J. D. (2001) *Marine Fisheries Ecology*. Blackwell: Oxford.

Loftus, E. F. (1996) *Eyewitness Testimony*. Harvard University Press: Cambridge.

Magin, U. (1996) St George without a dragon. In *Fortean Studies* **3,** (Moore, S. ed.). Fortean Tomes: London. pp. 223- 234.

Meurger M., and Gagnon, C. (1988) *Lake Monster Traditions: A Cross Cultural Analysis*. Fortean Tomes: London.

Naish, D. (2000) Where be monsters? *Fortean Times* **132,** 40 – 44.

Naish, D. (2010) Monsters of the deep. *Fortean Times* **262,** 36 – 37.

Paxton, C. G. M. (2002) In search of monsters? A defence of cryptozoology. *The Skeptic* 15(3) also in *Why Statues Weep* (eds. W.M. Grossman & C.C. French). Philosopher Press: London: Pp. 10 – 14.

Paxton, C. G. M. (2009) The plural of "anecdote" can be "data": statistical analysis of viewing distances in reports of unidentified giant marine animals 1758 – 2000. *Journal of Zoology* **279,** 381 – 387.

Paxton, C. G. M. (2010) The monster manual. *Fortean Times* **265,** 54 – 55.

Paxton, C. G. M., and Donovan, C., (in preparation) Presence only data in marine mammology: problems and interpretations.

Paxton, C. G. M., and Holland, R. (2005) Was Steenstrup right? A new interpretation of the 16th century sea monk of the Øresund. *Steenstrupia* **29,** 39 – 47.

Paxton, C. G. M., Knatterud, E., and Hedley, S. L. (2005) Cetaceans, sex & sea serpents: an analysis of the Egede accounts of a "Most Dreadful Monster" seen off the coast of Greenland in 1734. *Archives of Natural History* **32,** 1 – 9.

Pierce, S. K., Massey, S. E., Curtis, N. E., Smith, G. N., Olavarria, C., and Maugel, T. K. (2004) Microscopic, biochemical, and molecular characteristics of the Chilean blob and a comparison with the remains of other sea monsters: nothing but whales. *Biological Bulletin,* **206(3),** 125 – 133.

Prothero, D. R. (2007) *Evolution: What the Fossils Say and Why It Matters.* Columbia University Press: New York.

Radford, B. and Nickell, J. (2006) *Lake Monster Mysteries.* University of Kentucky Press: Lexington.

Regal, B. (2010) Where have all the werewolves gone? *Fortean Times* **259,** 46 – 49.

Reyna, V. F., and Brainerd, C. J. (2008) Numeracy, ratio bias, and denominator neglect in judgments of risk and probability. *Learning and Individual Differences.* **18 (1),** 89 – 107.

Roberts, D. L., and Solow, A. R. (2003). When did the dodo become extinct? *Nature* **426,** 245.
Roesch, B. S., and Moore, J. L. (2002) Cryptozoology. In *The Skeptic: Encyclopedia of Pseudoscience.* (Shermer M. (ed.) ABC-CLIO: Santa Barbara. Pg. 71 – 79.
Shermer, M. (1997). *Why People Believe Weird Things.* W.H. Freeman: New York. Pg. 48.
Shermer, M. (2003). Show me the body. *Scientific American* **288,** 27.
Schick, T., and Vaughan, L. (2002) *How to Think About Weird Things. Critical Thinking for a New Age.* McGraw Hill: New York.
Víkingsson, G. A., Pike, D. G., Desportes, G., Øien, N., Gunnlaugsson, Th., and Bloch, D. (2009) Distribution and abundance of fin whales (*Balaenoptera physalus*) in the northeast and central Atlantic as inferred from the North Atlantic Sightings Surveys 1987 - 2001. *NAMMCO Scientific Publications* **7,** 49 – 72.
Woodley, M. A., Naish, D., and Shanahan, H.P. (2008) How many extant pinniped species remain to be described. *Historical Biology* **20,4** 225 – 235.
Woodley, M.A. (2008) *In the Wake of Bernard Heuvelmans.* CFZ Press: Bideford.
Woodley, M.A. (2010) The case for cryptids. *Fortean Times,* 269,
Woodley, M.A., Naish, D., and McCormick, C.A. (in press) A baby sea-serpent no more: Reinterpreting Hagelund's juvenile 'cadborosaur' report. *Journal of Scientific Exploration.*

American Ibex Folklore

Chad Arment

There are, naturally, far fewer reports of mystery ungulates in North America than alleged lake monsters, anomalous primates, strange carnivores, and the like. Generally, there are certain biological, behavioral, and biogeographical characteristics that make herbivorous mammals less likely to remain unknown, but as every cryptozoology enthusiast should know, some of the most interesting zoological discoveries around the world in the last few decades have been previously unidentified ungulates.

I have only run across a handful of mystery ungulates in the U.S., and most were reported in the 1800s and early 1900s. Some, like the "fan-tailed deer," are likely cases of systematic confusion, as early hunters and trappers (and biologists) lumped or split species rather carelessly. A few others are less clear (e.g. Arment 2011). Presented here is one rather interesting case of a mystery animal that has a long ethnoknown history, but hasn't been discussed much (if at all) in the cryptozoological literature.

Stories of an "American ibex" come from several regions in western North America. The true ibexes, genus *Capra*, are ruminants found in Europe, Asia, and parts of northeastern Africa. They are closely related to the domestic goat. Male ibexes are well-known for their prominent long backward-curving ridged horns. This is the one feature that is usually pointed to, by witnesses of alleged American ibexes, as evidence that they were seeing an animal distinct from our known native caprids.

In North America we have the mountain goat (*Oreamnos americanus*), the bighorn sheep (*Ovis canadensis*), and the Dall sheep (*Ovis dalli*). We don't, however, have a recognized native species of ibex (*Capra*). Kurtén and Anderson (1980) noted that a partially mineralized caprid (classified as *Capra iowensis*) was discovered in an Iowa River sand bar, but it appears that this specimen has since been dismissed as a colonial-era feral domestic goat.

As long as there have been sightings, there have also been individuals (wildlife officers, biologists, hunters, etc.) pointing out the likelihood of misidentification. Particularly, they argue that younger male bighorns with incomplete horn curvature (or even old females) may present ibex-like appearances.

What follows is a chronological selection of historical passages from books and newspaper accounts noting cases of alleged ibexes and ibex misidentifications, first in Alaska, then from the Rocky Mountains (particularly from Idaho and Oregon), and finally from the southwestern states (Arizona and New Mexico).

ALASKA

> "It is worthy of note that the Indians report another animal unknown to naturalists, on the higher mountains of the mainland. It is said to resemble the mountain sheep and to have horns nearly as long but almost straight, like those of an ibex. Lieut. Emmons is confidant that these reports have a basis in fact." (Dall 1895)

> "In the window of the [Juneau] Post-office were displayed the horns and part of the frontal bone of the so-called Alaskan ibex, an animal affirmed by many to exist on the slopes of Mount St. Elias and the neighbouring mountains. I must say that, for my part, notwithstanding all the tales I have heard, I am very skeptical as to its existence, and the horns in question appeared to me to be suspiciously like

Alpine Ibex (Artur Tomasz Komorowski)

Male Bighorn Sheep (Bob Loblaw)

Young Bighorn Sheep (Jonathan Allers)

Rocky Mountain Goat (Henryk Sadura)

those of the common or garden goat. The proprietor of the shop could give no precise account as to how, when, or where they were procured. All he knew was that the man he bought them from told him he had not shot them himself, but that he in turn had bought them from someone who assured him they had belonged to the veritable Alaskan ibex, and had come from the slopes of Mount St. Elias.

"These horns did not in the least resemble any ibex horns I have ever seen, either Indian, Caucasian, or Spanish; but might at first sight have been mistaken for a very small pair of markhor horns; the twist of the spiral was, however, inwards instead of outwards, which confirmed my suspicion, stated above, that they had originally adorned the head of a common domestic goat, and not a very big one at that. It seems unlikely that such a species should exist in Alaska, separated by nearly half the circumference of the globe from the rest of his relations; but, of course, when one remembers the case of the American wapiti (*Cervus canadensis*) and the Altai wapiti (*Cervus eustephanus*), one must not take it for granted that a species cannot exist in a given locality, merely because many thousands of leagues of land and water intervene between it and the *habitat* of its nearest allied congeners." (Cane 1903)

"I feel some diffidence in approaching the following subject, but since most sportsmen who have visited the western coast of Alaska have probably heard similar yarns, I will quote a few statements which I gathered in regard to the fabled ibex of Alaska.

"It is a popular theory amongst many prospectors and others who have been far into the mountains of the interior that an animal exists there which is neither a sheep nor a mountain goat. I had long talks on several occasions with one of the oldest and most trustworthy residents, who has lived for many years in the head of Cook's Inlet. He declared to me that he had seen some horns brought out by natives from the mountains some miles inland which were like the horns of the *Ovis dalli* ewe, but were more than twice the length of the sheep's horns and also larger. He said he imagined they belonged to some kind of ibex.

"Mr. J. Folstad, the owner of the schooner *Alice*, in which we sailed along the coast of the Alaska Peninsula, who was a Norwegian engineer, and had travelled over a great part of the interior of Alaska, makes the following statement.

"During the first great rushes to Dawson in 1897 and 1898 he was in the town of Dawson. There he saw, hanging outside a store (the owner of which sold meat such as moose, sheep, etc.), two animals with long horns which were unknown to him. On inquiry at the store he was told that numbers of people had been to look at the animals, and no one knew what to call them. But a sporting English doctor in the town, hearing of the episode, had been to examine them, and pronounced the beasts to be undoubtedly ibex. The only part of this story which I cannot understand is why this doctor did not obtain the horns of these beasts, and why the sporting world heard no account of his discovery.

"On board the ss. *Bertha*, going down to Seattle in November 1903, we heard that there were two prospectors on board who had seen specimens of the fabled ibex. Mr. Vander Byl and I went to interview these men. They both described the animals as being darker than sheep, with shaggy hair underneath the throat and belly, horns brown in colour and curving backwards, about 30 inches long. One man claimed to have seen three of four dead ones killed by prospectors, and he said he had seen them in the mountains, sometimes in company with sheep, but that they were not so numerous as the sheep. We produced some photos of dead ibex killed in the Altai Mountains, and asked if he had ever seen a similar animal. He at once replied, 'Why, that is the same beast, I guess.' He declared these animals could be seen near the head-waters of a tributary on the east side of the Copper River.

"I may mention that early in 1904 two sub-species of wild goat from British Columbia and Montana were described by Dr. J. A. Allen, but these, of course, have nothing to do with the reputed ibex.

"Now the only explanation of the riddle appears to me to be this. Either there is some animal in Alaska as yet unknown to naturalists and sportsmen, or these animals are descendants of domestic goats which have escaped from old Russian or native settlements in former times. The latter theory is, I think, borne out by the so-called Mount St. Elias ibex. Colonel Cane, in his book *Summer and Fall in Western Alaska*, mentions a pair of horns which may be seen in the Post Office at Juneau, and are called the horns of a Mount St. Elias ibex. It is claimed that the animal was killed on the slopes of Mount St. Elias. I have seen and examined these horns, and entirely agree with Colonel Cane's remark that 'they might at first sight have been taken for a very small pair of markhor horns; the twist of the spiral was, however, inwards instead of outwards.'

"On mature consideration, these horns certainly appear to have belonged to some form of a domestic goat, although I cannot say that I have actually seen any goat with a head closely resembling the one in question.

"Doubtless some enterprising sportsman will ere long make an expedition up into the Copper River country, where there is also reported to be a particularly savage species of brown bear called locally the 'bald-faced bear.'

"I have heard tales of these bears lying in wait for men on the trails, and killing them without being previously wounded.

"Any one who listens to half the yarns regarding big game in Alaska is likely to be led many merry dances on a fool's errand, but the stories regarding the ibex come from so many different parts of the country, and are all so very similar, that it leads one to believe that there really must be some 'fire' behind the smoke." (Radclyffe 1904)

"Readers of *Recreation* know that the world has frequently been startled by reports of the discovery, in various places in the West, of an ibex. We found one near one of our camps, among the high tops. That is, we found what many a man would have called an ibex, without making a thorough investigation. Here is a picture of him. However, on close examination he proved to be simply 2 sprouts that had grown up from the root of a fallen tree and died and shed their leaves. It happened that the root of the tree had been burned and a remaining bit of charcoal formed what appeared to be the animal's right eye. A piece of another sprout that had been broken off furnished a good imitation of an ear, and, viewed through the brush, the outfit looked very like a real ibex.

"Many a so-called hunter would have plunked a bullet at such an apparition, and then have gone to camp and told the boys how he had shot at an ibex, and hit it between the eyes, but that when he went up to it, it proved to be only a root of an old tree. There are other hunters who would have plunked a bullet 4 feet to one side of it, or over it, or under it, and would still have told the other fellows how they hit it between the eyes. I know certain tenderfeet who would have fled from so formidable a looking beast, at sight, and rushed into camp wildly excited and told the boys between gasps, how they had seen an ibex, and that it dashed into the brush and escaped before they could get a shot at it.

"This picture shows about as good a specimen of the real ibex as has probably ever been found on this continent; yet a well known British Columbia sportsman told me 2 months ago that he firmly believed there were plenty of ibexes in a certain remote part of Alaska at that time. He said he had been assured of this fact by men who had seen them and whose word he could not doubt. I asked him how he accounted for the fact that though while men and Indians had hunted in Alaska a hundred years, not a single head or skin of an ibex had ever been brought out? He shook his head, but said he still thought there must be living specimens of this animal up there." (Shields 1904)

"Many an old-timer in Alaska will tell the visitor that a species of ibex different from both sheep and goat inhabits the territory. When such stories are sincere they undoubtedly refer to young goats or to either the young or the female of the white sheep, for no true ibex is native to any part of North America." (Osgood 1909)

Huge Lituya Bay Goat Believed to Be Ibex

Juneau, Alaska, *Daily Alaska Dispatch*, February 17, 1918

Does the ibex—the long unaccounted for parents of all the goats of the North American continent—have its home, along with so many other mysteries, in the Alaskan wilds? Does the Lituya Bay district, within a short distance of the capital city of the Territory, have the distinction of sheltering the herds of these mysteriously disappeared progenitors of the goat race in this half of the Northern Hemisphere?

All dependable evidence would indicate that such is the case and that for the scientist whose delight is the ferreting out of such problems, a new occasion of pleasure is still in store.

For the men who have seen and hunted this strange animal, do not pretend to be learned in the lore of the naturalist, but they do maintain that the great goat which they have frequently encountered in the Lituya Bay region is not an ordinary goat by any means, and that it does fit the descriptions of the rare ibex.

There is a well-accepted theory among naturalists that four Alaskan islands, Admiralty, Krupeanoff, Kodiak and

Graham, supplied the grizzly bear population for the entire Rocky mountain range of North America. From the brown bear stock on these islands the bruin lords of the Rockies are believed to have descended.

It is not altogether beyond belief that, as one particular variety of ibex, found in Asia Minor, is supposed to be the parents of the domestic goats of the Old World, the Lituya Bay mammal tribe provided the goat population for the New World.

Several varieties of ibex, or chamois, are found in Europe and Asia. The Pyrenees, Alps, Altai, Himalayas and Caucasus mountains have scanty herds of them, differing mainly in the form of horns. All, however, have long recurved horns, transversely ridged in front.

The Alpine ibex, called the bouquetin, is the best known, although it is now rare except where protected.

E. Doyle of Lituya Bay is one of the numerous hunters and woodsmen who have noted the strange species of goat which is believed to be the ibex. Robert Barclay, who has killed many of them, supports his contentions that the animal in question has marked characteristics differentiating it from the common mountain goat.

The female of the Lituya Bay ibex have four teats or dugs, whereas the mountain goat has but two. The horns have a tendency to curve around more like those of a mountain sheep, while the true goat has shorter horns. The fleece or mohair from these ibex is extremely long, fine, and silky. It is much finer than that of the goat, finer even than the mohair taken from a domesticated angora goat. This fleece is so heavy that a bullet from a high power rifle has been known to lodge only a short distance under the skin, the cushion formed by the heavy fleece having stopped the force of the bullet.

In weight they are from thirty to forty per cent heavier than the goat, animals dressing five hundred pounds being frequently taken. They are browsers the same as the goat, and while they do not exist in countless numbers, they may often be seen on the hill sides above timber line in herds of a dozen or two. When the heavy snow storms of winter set in, they are forced down, sometimes even to the ocean beach.

The enormous size of these animals has been widely commented upon. Some men describe them as being "as big as camels." Like the European chamois, they are impressive looking creatures, far exceeding in stature and build any ordinary wild mountain goat.

So far as known, they are to be found only from Laprue Glacier westward to the Alsek river, and inland for less than a hundred miles.

Other reports from the Yukon district just back of this section, lends further support to belief in the existence of the ibex in Alaska.

And fifty miles toward the Lituya Bay section from White Horse is a stream called Ibex river.

[While the author was hunting Dall sheep:] "While still within range of a long shot, they stopped and looked back, as if wondering why the leader did not follow. Then they ran upward in a series of jumps for a hundred feet and stopped again. This method of ascending was continued until they reached the crest, when again I had before me that wonderful sight of the mountain ram walking on the sky-line. I had noticed particularly their attitudes. The large one seemed to be all elasticity. His head was held straight up, his neck swelled out, his back straight, his legs rigid. He reminded me of a strutting cock. Thus appears the mountain ram under excitement. But I made an interesting discovery. The other ram was three or four years old and his horns curving behind did not curl upward. When he reached the skyline and came under a certain angle with the sun, I beheld the famous 'ibex' often reported to have been seen on the far northern summits of Alaska. The horns appeared to be magnified to a long sweep of three or four feet, curving behind exactly like those of an ibex. So complete was the deception, that had I not been certain of the animal, I would have been deluded into the belief that an ibex stood before me. I have many times witnessed similar delusions and have thus learned to give more credit to the good faith of those who report having seen an ibex. Also, I know that they have not been so fortunate as to have had experiences similar to my own." (Sheldon 1919)

"Next day it continued raining, so the [story] contest was resumed, lasting all that day and far into the night. Shorty told of once capturing a goat alive in Alaska, and said they were so tame and plentiful that it would be no trick at all to repeat the performance on this trip. Cap said he had seen the rabbits so thick in that country that they ate off all the vegetation—in fact, these rabbits were so numerous that finally they had no feed whatever, so they ate themselves. Billy Wooden told of killing an ibex in Alaska, describing it as a counterpart of the goat except that the front feet were large and the horns were twisted, containing ridges that ran in spiral fashion around the horn, as in some of the European species.

"I was curiously interested in the ibex story, especially as I had heard from other sources of these animals having existed there. One man who vouches for their presence at one time in Alaska is ex-Representative James Wickersham, of Fairbanks, with whom I conversed on the subject.

"However, Judge Wickersham, I believe, received his impressions more from what he read in Gen. T. A. Allen's book, 'Government Report on the Copper River (Alaska) Exploring Expedition of 1886,' than from any personal experience that he has had with the supposed animals. I have a copy of General Allen's book, and publish herewith an extract from it covering the subject, as follows:

"'Whether the big-horn mountain sheep, ovis canadensis, exists in Alaska I am unable to say, but I desire to add also a new geographical race of the same. The animal in question is called by the natives tebay, and this name I leave unchanged until a specimen will have been carried out of the territory. We killed several of these animals, one of which, a ram, had horns twenty inches long and nearly straight. Their structure was similar to that of the bighorn, but the curvature was very slight. This ram was killed on a very high point, such a place as is usually sought by them, and in its fall was sadly mangled. The head of the tebay is much like that of a Southdown sheep, the muzzle much less pointed than in Nelson's big-horn. The hair is of a uniform white—in fact, nearly equal to his snow surroundings in color, and is nearly as easily broken as that of the antelope. Next to the skin is a very fine, short wool, which is very strong. In size the tebay is probably an equal of its relative, the bighorn. I saw a spoon made from the horn of one that measured twenty-six inches in length and five inches across the bowl. We were informed that some had much larger horns than the one that furnished material for this spoon. This, like most statements of natives, is questionable. The large ram and one other were killed on the most northerly tributary of the Chittistone River. The natives informed us that small tebay could be killed a few miles below the junction of the Chittistone, a fact we doubted, and hence chose to allow them the use of our carbines. They passed the night on the mountains north of the Chitina River, and returned with four small ones that would weigh when dressed probably sixty-five pounds. The heads were left on the mountains, but the bodies brought in seemed identical with those obtained on the Chittistone River. Why only small ones should be found at this place in the latter part of April I cannot say; yet the mountains here were not so high as farther to the east, where the large ones had been killed. The last of these animals seen or heard of by us were near the headwaters of Copper River, on the divide between it and the Tanana River.'

"At this late day, of course it seems odd to read of a doubt cast at the habitat of the ovis canadensis, as shown herein by General Allen, but when one reflects that his book was written about thirty-five years ago, it is not amazing. It is amusing to note the two very distinct animals described respectively by Billy Wooden and General Allen. Billy Wooden's animal of mystery was distinctly a goat, except for the horn and front hoof formation, while General Allen's was a sheep. There could, of course, be no connection between the two forms, according to the descriptions given. Naturally, when we hear of such reports, the first thing that enters our mind is that no hunter has ever been able to secure and preserve one of the skins, and secondly, that none of these specimens has ever reached any of the many natural history institutes of our country that would be so very anxious to secure them at a substantial cost. I believe I can solve the Allen myth by suggesting that it might be a young mountain sheep ram or an old female, with slightly curved horns. But Billy Wooden's ibex has simply got my 'goat,' for I cannot fathom it. Rumors of ibexes having been seen in the States are very old. Other unnatural forms of wild life have also been reported, but when run down they have usually turned out to be about as authentic as the stories of the philaloo bird and the side-hill gouger." (McGuire 1921)

See Ibex in Alaska

Freeport, Illinois, *Journal-Standard*, September 10, 1924

Tanana, Alaska.—A species of ibex—identical with the Old World ibex—has been sighted by hunters in the Noatak River region. This wild country abounds in both sheep and mountain goats, but the hunters insist that they have discovered the true ibex.

"On the other side of the Bering Strait, the Eskimos of Alaska also have legends, myths, and rituals concerning the animal of the snows. In pictures etched on walrus ivory, one can see strange quadrupeds with long horns and short, stocky legs, which look like ibex when their 'horns' point backward, or reindeer when they point forward. But there are no ibex in Alaska, and these figures, which some ethnologists consider mythical goatlike animals, are probably firsthand renderings of mammoths, based on the shape of the animal's body, deformed by ice." (Cohen 2002)

ROCKY MOUNTAINS

Owyhee, Idaho, *Avalanche*, June 17, 1893

Frank Leply saw four ibex on a bluff of rocks above Louse creek one day this week. He is familiar with these animals, having killed numbers of them on Burnt river, Oregon, and says he could not have been mistaken; but to make sure of it he sh—got one of them. These are the first of these, apparently a connecting link between mountain sheep and goats, we have ever heard of in Owyhee.—*Nugget*.

"Is There an Ibex in this Country?

"For 20 years past I have been hearing rumors and reading newspaper accounts of the appearance of so-called ibex in Idaho, Montana, Oregon, Washington or British Columbia. I have always regarded these stories as pipe dreams and have run down several of them. In each case where it had been claimed that the ibex had been killed the specimen proved on examination either a female mountain sheep or a white goat.

"The region in which it is claimed the ibex has been found has been worked over by fur traders and trappers for 200 years, and if there were such an animal there some one of those trappers, or some sportsman or naturalist, would surely have taken a specimen and it would have found its way into some museum.

"Here is a letter that reads more like fact, more like the report of a careful observer, than any ibex story that has ever come to my notice:

"Tacoma, Wash., Nov. 12, 1901.

"In May, 1900, I was staying in the Crow's Nest pass, Canadian Rockies. One evening while going along the railroad track 2 animals came down from the mountain side to the track. They would have weighed about 160 to 180 pounds each. The color was ashy gray. The hair on the back was about 6 inches long and stood up, forming a black line from the back of the head to the tail. The under side was a little lighter color. Legs were of medium length. The horns were about 3 feet long, beautifully black, curved back and a little outward. I was within 15 yards of them and had a good

IS IT AN IBEX?

look at them. I believe they were ibex. I send you a drawing of the head of one, made while looking at them. I hunted for them a week or so after that, but never saw them again.

"A rancher in Alberta told me that in old times when he lived in Oregon he saw several of these animals, which he called antelope. He recognized the drawing instantly as of the same animal.

"John F. Almon.

"Could these have been the descendants of some domestic goats that were liberated years ago? Or is there really a species of ibex indigenous to the Northwest? Who can tell?—Editor." (Shields 1902)

"Legendary Distribution.

"The writer has carefully traced out the legends regarding the occurrence of goat in Colorado, Utah, and California. There are persistent stories about the existence of white goat in Colorado, which, when investigated seem to have their origin in some domestic goat which are known to have escaped from captivity. It is, however, a certainty that *Oreamnos* has not existed in Colorado since the arrival of the white man, and there is no proof of its previous existence there. This statement is made after a full examination of the evidence.

"The purpose of this paper has been to gather and summarize the known facts about this interesting animal and it has been necessary to discard a large amount of data contained in the literature of the subject. Statements by certain writers regarding the existence of the goat in Wyoming, Colorado, California, and even New Mexico, are extremely misleading. It is positively known that no goat have ever existed on Mt. Shasta, although this mountain has been a favorite locality for stories about mountain goat and the mythical ibex. The origin of these fables is easily traced to the former existence on Mt. Shasta of mountain sheep, the horns and bones of which are still occasionally found there. The straight horns of the mountain sheep ewe are probably responsible for most of these legends. It is bad enough to suggest the occurrence of goat on Mt. Shasta, but it is utterly absurd to assert their existence on Mt. Whitney, 300 miles farther south, and it is still worse to include in the range of the goat New Mexico or the barren coast mountains of southern California.

"The above examples will suffice to show the loose manner in which this subject has been treated by writers who have not sifted the evidence sufficiently.

"Within the United States the mountain goat is only found in Idaho, western Montana, Washington, and Oregon. There is no evidence whatever of the white goat having existed in Wyoming. In examining the rumors respecting the occurrence of goat one must remember that only a few years ago very little was known about this animal, and few people had seen it. In the south, escaped domestic goat and old mountain sheep ewes with bleached coats and straight horns, have probably been the basis of many such stories. In some places such animals have been mistaken for white goat and elsewhere, notably in Alaska, for the legendary ibex. Until the discovery and description of Dall's white sheep, in 1884, all white animals in the north were called goat and white mountain sheep meat is sold to-day in Dawson City restaurants under that name." (Grant 1904)

Ibex in Oregon

Boise, Idaho, *Idaho Statesman*, August 15, 1908

Baker City, Ore., Aug. 14.—Deputy Game Warden W. R. Parker of this city, accompanied by a party of hunters has just returned from Main Eagle, northeast of Baker, where they killed three ibex. This is something extraordinary for, according to reports, there are few ibex in the United States and they never have been seen outside of eastern Oregon and Idaho. The game warden was well pleased with the game and it has attracted an unlimited amount of attention here.

Save Ibex His Aim

Portland, Oregon, *Oregonian*, November 22, 1909

Advocating that the State of Oregon or the Federal Government should surround Mount Eagle in the Cornucopia Range with a game preserve that the only herd of ibex known to exist on the American continent may be saved, is one of the purposes of the present visit to Portland by W. R. Parker, of Baker City.

Mr. Parker astonished big game hunters a few years ago when he announced that he had found a bunch of the rare animals in an almost inaccessible part of Mount Eagle. Having been engaged in hunting the big game of Eastern Oregon for many years and possessing superior knowledge of the state's mountains, his statement was not openly scouted, but there were many throughout the country who wanted proof of the existence of the ibex. In company with two brothers, Mr. Parker returned to the mountains and killed a buck, doe and fawn. He brought them out and had them mounted, as well as photographed before removing the hides.

Known as "Bill" Parker among the crack shots and hunters of the state, he recounts many stories of his hunting days, and is always surrounded by a group of attentive listeners when he reaches the Hotel Perkins, his headquarters. Last night he said:

"We went in to the Cornucopia Mountains in 1906 after a bunch of bear, having been in the habit of following that sport every Spring. We first located three of the animals, but later my brother found about 26 in all. The ibex have been unmolested, so far as we know, since the Fall of that year, when I killed the specimens referred to for scientific purposes. We have always refused to guide hunters to their feeding grounds, although we acted as guides for many years.

"That country is now filling up with settlers, and in my opinion it will be only a short time until somebody runs on to the haunts of the little bunch of ibex, the only herd known to exist in America. It would seem advisable that some protection should be accorded them, and we stand ready to give the proper officials all the information we have gathered. If it is not done they will probably disappear, as have the mountain sheep in some parts of the mountains."

Mr. Parker is an expert with a revolver, rifle or shotgun, one of his favorite tricks being the breaking of marbles with rifle or pistol bullets as they are thrown in the air. He shoots the automatic guns and carries several beautifully engraved with his name and address and a number of hunting scenes.

"An American Ibex

"Idaho enjoys the proud distinction perhaps of being the only state in the Union that has found it necessary to enact a law protecting ibex. Not that there is or ever was one of these animals in the state, but some time back in the past a legislator opined there was and got his opinions enacted into law. Each succeeding legislature has preserved the tradition and solemnly re-enacted the law. So it stands today, a monument to the game preservative zeal of the state. The hunter is solemnly warned, under severe pains and penalties, that should he encounter an ibex making a meal off the cobblestones of some lofty mountain peak, he is to stay his lethal arm and permit the animal to go frisking away unscathed. The fact that no animal of the genus *Capra* is native to the New World seems never to have found lodgment in the brains of our law makers; to them the ibex is an actual fact.

"The above law is mentioned for the guidance of visiting sportsmen whom the writer wishes to have enjoy their sojourn among our mountains and streams unmolested by the attentions of the game warden. The sportsman is warned not to interfere with the peace and prosperity of our ibex; Idaho has seen fit to go into the ibex business and we object to having our supply killed off. Of course, the possibility of your ever encountering one of these animals is about nothing to infinity, but it is as well for you to be warned beforehand.

"The writer encountered this strange statute several years ago when serving a sentence in the legislature, and when the Fish and Game Committee reported its bill was somewhat surprised to find the clause occupying a prominent place in the measure. He humbly asked the assembled solons if any of them had ever killed an ibex in Idaho, or if they knew of any person who had been so fortunate. The inquiry disclosed the fact that no one present had ever killed the animal, nor could any lawmaker at that moment remember a man who had done so, but inasmuch as it was written into law that ibex should be held sacred it stood to reason that some person had been guilty of the felony and it was just as well to awe that fellow and others like him with the majesty of the law against a repetition of the offense.

"It required several years of diligent search to trace the origin for the belief that ibex frequented our state. Like many similar mistakes in natural history, the matter had its origin in ignorance of what constitutes a species, or rather a genus of a species.

"About twenty-five years ago some hunters brought into the city of Boise from somewhere in the Salmon River mountains the skin of a goat which differed materially from the ordinary Rocky Mountain White Goat (*Oreamnus montanus*) (sic). At least people who saw the skin said it differed, and after due debate it was decided that the animal must be an ibex. The animal was white, somewhat taller and more slender in form than the ordinary goat, the horns were much longer, having both a backward and lateral curve, the points flattened instead of sharp, and it did not possess the hump of the mountain goat. These were the principal distinctive marks and upon these a classification was made.

"Other hunters reported similar goats in various parts of the state, generally from the higher ranges. These kept together in small bands and were exceedingly wary. As the years went on these ibex (?) became scarcer until, like the buffalo, only a few remained. Then came the wise legislator and insisted that the law's restraining hand should be laid upon people bent on destroying our ibex herd.

"While not particularly relevant to the subject, the writer is constrained to relate an anecdote which will serve to illustrate how easy it might be for persons not trained in anatomical matters to mistake a simple type variation of a genus for an entirely new genus. He strayed into Withers Brothers in the city of Spokane, Wash., and was confronted by a beautifully mounted head upon the wall. At first glance the head appeared to be that of an African antelope, the next glance and it appeared to be a white-tail deer head with the horns of a goat.

"'What sort of monstrosity is that?' he asked.

"'That,' Jack Withers replied, 'is an Australian Koodoo, probably the only head of its kind on exhibition in the United States.'

"Jack made the mistake of getting his koodoo in the wrong country or his hoax might have 'stuck.'

"A close inspection failed to disclose the identity of the animal and the writer was compelled to call for help.

"'What in blazes is that, Jack?' Jack laughed. 'You are about the only man that has looked at that head who has not been satisfied with my explanation, and some of them have offered me big sums for the rarity. To tell you the exact truth, that is a white-tailed doe scalp with a pair of domestic goat horns fitted to it.'

"Dr. Hornaday, who has examined the so-called ibex of Idaho is of the opinion that it is nothing but *Oreamnus* with minor type variations. He found similar characteristics existing in the goats of the Kootenia mountain region in British Columbia, and elsewhere along the Pacific coast in the northwest. Other careful zoologists have agreed that the animal is not a *Capra* nor does it belong to the subfamily *Caprinae* but to the subfamily *Rupicaprinae* or short-horned goats." (Anonymous 1912)

Protecting the Ibex

Oakland, California, *Tribune*, December 18, 1912

Idaho enjoys the proud distinction perhaps of being the only state in the Union that has found it necessary to enact a law protecting Ibex. Not that there is or ever was one of these animals in the state, but some time back in the past a legislator opined there was and got his opinions enacted into law. Each succeeding legislature has preserved the tradition and solemnly re-enacted the law. So it stands today, a monument to the game preservative zeal of the state. The hunter is solemnly warned, under severe pains and penalties, that should he encounter an ibex making a meal off the cobblestones of some lofty mountain peak he is to stay his lethal arm and permit the animal to go frisking away unscathed. The fact that no animal of genus capra is native to the New World seems never to have found lodgment in the brains of our law makers, to them the ibex is an actual fact.

It required several years of diligent search to trace the origin for the belief that ibex frequented our state. Like many similar mistakes in natural history, the matter had its origin in ignorance of what constitutes a species, or rather a genus of species.

About 25 years ago some hunters brought into the city of Boise from some where in the Salmon River mountains, the skin of a goat which differed materially from the ordinary Rocky Mountain white goat (oreamnus montanus). At least, people who saw the skin said it differed, and, after due debate it was decided that the animal must be an ibex. The animal was white, somewhat taller and more slender in form than the ordinary goat, the horns were much longer, having both a backward and lateral curve, the points flattened instead of sharp, and it did not possess the hump of the mountain goat. These were the principal distinctive marks, and upon these a classification was made.

Other hunters reported similar goats in various parts of the state, generally from the higher ranges. These kept together in small bands, and were exceedingly wary. As the years went on these ibex (?) became scarcer, until, like the buffalo, only a few remained. Then came the wise legislator and insisted that the law's restraining hand should be laid upon people bent on destroying our ibex herd —*Outing*

Idaho Hunter Kills Ibex

Boise, Idaho, *Idaho Statesman*, October 1, 1913

Salmon, Ida.—An ibex was killed last week by Lee Ramey on the basaltic crags between Singiser and Rabbitfoot. This is the first ibex known to have been taken in that part of Lemhi county. There is a difference of opinion among authorities as to the existence of ibex in Idaho, and a writer in an eastern magazine recently made sport of Idaho legislators for including the ibex in the list of game animals in this state. Last winter a small band of six ibex was stalked near Boyle creek, in this county, and one of the animals was shot by the hunter, who did not know what he had.

District Forester Grandjean Hoots at Ibex Story

Boise, Idaho, *Idaho Statesman*, October 3, 1913

District Forester Grandjean takes exception to a recent news item from Salmon stating that Ibex had been killed in that vicinity. He maintains that the ibex family has no representatives on the American continent, being found only in Europe, Asia, and Africa. He has often noticed statements in the state press to the effect that Ibex have been found within the confined of the state of Idaho, but these assertions have never been proven by producing the remains of any such animal.

Several years ago when a statement appeared that ibex had been discovered in this state near the headwaters of the Salmon river and in the Sawtooth and Salmon river mountains Mr. Grandjean took this matter up with the Smithsonian Institute, and the institute informed him that they often received specimens for identification which the senders claimed to be ibex, but that in all instances the animals had proven to be of the Rocky Mountain goat or the Rocky Mountain sheep family.

How Errors Are Made

"That a large animal like the ibex could exist, even in the most remote part of Idaho without ever being identified or without the horns or head of such an animal ever being submitted to the proper authorities, seems highly improbable," said Mr. Grandjean. "People who are not well posted on the animal life of this country are apt to get the impression when viewing a female mountain sheep or a band of yearling rams that they have discovered some strange animal."

Chance to Earn $500

Mr. Grandjean says that several years ago he made an offer of $500 to the party who would bring in any specimen of the ibex family which had been found in this state. Several hunters at that time claimed to have a specimen but no one ever produced any proof or claimed the reward. He says this offer has been withdrawn, but feels sure that a substantial reward from the Smithsonian Institute awaits the hunters who will furnish proof that a specimen of the ibex family, in its wild stage, has been found in North America.

"The Irrepressible American 'Ibex'

"Throughout the Rocky Mountain region from El Paso to Dawson City belief in the existence of an undiscovered American Ibex springs eternal in the human mind. Again and again has the creature been seen and reported, with positiveness and particularity. From the State of Washington, one man sent a very good drawing of its head and horns, and from Colorado came a photograph, an admirable description, and measurements, of a specimen which had actually been shot and mounted. Two really distinguished sportsmen of our acquaintance were with some difficulty convinced that a journey in pursuit of the horned mystery would be a waste of time.

"The spirit of investigation which prompts the pursuit of a mysterious animal, is highly commendable.

"Without it the scientific world would lose much. At the same time, it is unfortunate that all Rocky Mountain hunters can not know that there really is not the faintest probability of the discovery in America of any living representative of the genus Capra, and that it is useless to pursue the phantom 'Ibex' of the West.

"The specimen shot in Colorado, and submitted to us, was a domestic goat, presumably of Spanish breed, that had escaped from captivity and become wild and self-supporting. Such animals account for some of the 'ibexes' that have been observed. A pair of horns and a pelt recently sent to

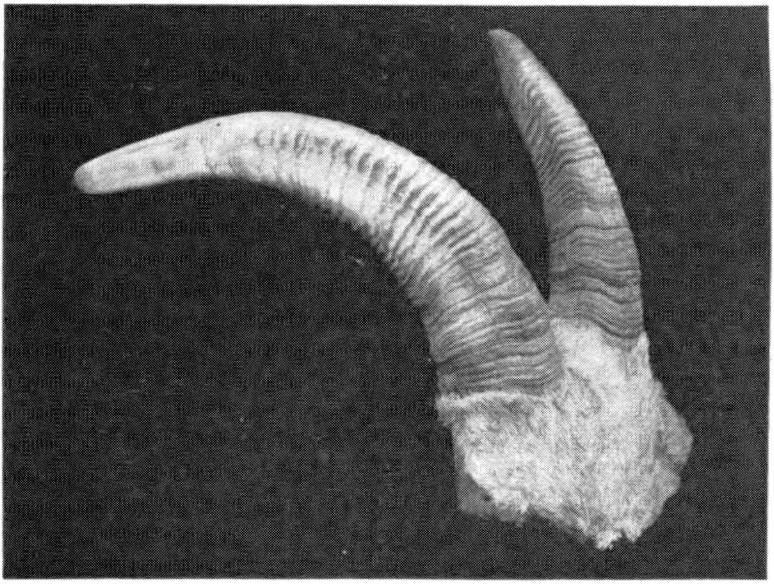

ROCKY MOUNTAIN SHEEP HORNS.

us by Dr. D. T. MacDougal, from the Desert Botanical Laboratory at Tucson, Arizona, illustrate another source of honest belief in the existence of an American Ibex. Dr. MacDougal, who is himself a keen naturalist, had no difficulty in naming at sight the species which these specimens represent, but he kindly elected to afford us another practical demonstration of an 'Ibex' story reduced to its lowest terms. The animal shot as an 'Ibex' in the Santa Catalina Mountains of Arizona proves to be a big-horn mountain sheep, female (*Ovis canadensis*), about four years old. As in all horns of female mountain sheep, these describe only a quarter of a circle, and in their lack of curvature they are slightly goat-like.

"Beyond doubt, the many 'Ibex' stories and queries that have so frequently arisen during the past fifteen years, originated in honestly made but wholly erroneous observations of domestic goats running wild, of mountain sheep ewes, whose horns are short and rather straight, and of young mountain sheep rams.

"In this connection, it may also be noted that in many instances female white mountain sheep seen at a distance have been mistakenly identified as mountain goats." (Hornaday 1915)

"The Rocky Mountain Goat, or White Goat, is the only American representative of the numerous species of wild goats, ibexes and other goat-like animals so numerous throughout the Old World from Japan to India, southern Europe and northern Africa. Thus far with but one exception all the rumors of 'ibex' that have come from Wyoming, Colorado, Montana and British Columbia have proven entirely without foundation. In one case a Colorado hunter discovered a small band of once-tame goats running wild and reported it to *Recreation* magazine, with a photograph of a mounted specimen. While it is possible that a genuine *Capra* may yet be found inhabiting some unexplored region, like the Romanzoff Mountains, such an occurrence is very improbable." (Hornaday 1919)

Idaho Seeks Famed Ibex

Portland, Oregon, *Oregonian*, October 29, 1919

Boise, Idaho, Oct. 28.—(Special.)—The state fish and game department is seeking the famed ibex in Idaho, now almost extinct. Reports coming to that department indicate the animals are to be found in mountain fastnesses. Otto M. Jones, state game warden, has decided to investigate. He has gone into central Idaho on that mission and will have to follow trails over which few human beings have passed in order to reach the supposed hunting ground of this animal.

This animal is supposed to be found only in the Asiatic mountains. It is very similar in appearance to a mountain goat. The game warden will also investigate sheep ranges to determine if the sheep are interfering with the deer.

Idaho "Ibex" Just Mountain Sheep; Jones Discovers

Boise, Idaho, *Idaho Statesman*, November 8, 1919

The next Idaho rancher, sheep man or hunter who kills an "ibex" will face a fine of $100 and costs, ruled Otto M. Jones, state game warden, Friday.

"There ain't no such animal," the warden declared after reviewing his two weeks' search in the middle fork country of the Salmon river hunting for clews and examining alleged specimens. "The ibex was never found in Idaho."

Two alleged samples of the animal were viewed by the warden on his hunt and both were found to be ewe mountain sheep, a forbidden animal to the hunter. Their short slightly curved horns and a suggestion of corrugations upon them had misled the hunters.

The game warden found that residents of the middle fork country had mistaken mountain sheep for ibex many times, particularly when young bucks were found with horns not yet grown into the characteristic sweeping curve and ewes with similar head dress.

Sometimes the animals had been called "chamois" and sometimes had been set down for an entirely new breed of mountain animal.

The warden, who pressed into the mountains to such distances that he was in great danger of being snowbound, who froze part of one foot in his expedition and who battled for a week to get back to civilization, convinced himself by the trip that all the ever-recurring ibex rumors are false.

The next hunter who thinks he has found a new animal, an ibex or a chamois, he said, is bound to find instead, should he kill the animal, a justice and a stiff fine. Ibex hunters will hunt at their own peril.

Idaho Protects Ibex

Pittsburgh, Pennsylvania, *Press*, February 6, 1920

The state game warden in Idaho warned the ranchers that hereafter the slaying of an ibex would mean $100 fine with costs. Now an ibex, outside of the ninth page of the nursery alphabet book, was an alien animal in those parts, so with the keenest zest every one who owned a gun hastened out to find such a creature. Several sample ibices were bagged and put on view, only to be pronounced mountain sheep, which, before their horns were grown, the more cultured Idahoans prefer to call chamois. But bent upon an ibex, for purposes of comparison, the warden pressed far into the mountains in his quest. He just escaped being permanently snowbound, froze a foot and battled with the storms for a week to get back to civilization, where arrived, he announced that all ibex rumors are false, and that any one who thinks he has shot one had better conceal it.

Woods and Waters, by Pierre Pulling

Pocatello, Idaho, *State Journal*, September 21, 1952

This column in the past has referred to the Nevada ibex myth. There is an "Ibex Pass" near the California-Nevada boundary, and I have talked to several Nevada mountaineers who claimed to have seen herds of ibex. Nevada had a law protecting the ibex years ago. The true ibex is a European goat, and nearly extinct.

There is no proof as to what these fellows saw, or how sober they were when they saw them. Further, elective bodies can pass anything, like the club man who was elected cuspidor, and proud of it. I think they saw bighorn ewes.

Milt Williams (1977) of Idaho Fish and Game suggested that early reports of ibex in Idaho were the mistaken identity of bighorn sheep ewes and their "relatively thin, slightly curved horns." He did note that in a booklet by Adelaide Hawes (*The Valley of Tall Grass*), she noted an 1883 newspaper article telling of "a group on a pleasure trip into the mountains of Owyhee County having killed forty-five deer, seven antelope and three ibex."

SOUTHWEST

Syracuse, New York, *Daily Standard*, August 12, 1895

An ibex was killed in the Superstition mountains, Maricopa county, Ariz., recently. It is said that only two others of these animals have ever been killed in the United States. They were found in Oregon.

Ibex at Willow Creek

Tucson, Arizona, *Daily Citizen*, August 13, 1913

Prescott, Ariz., Aug. 12—Residents of Willow Creek report seeing an ibex near that point. This rare animal has rarely been known in Arizona. It has been reported on an average of twice a week. There is a $500 fine imposed by the state game law for killing the ibex.

Ibex Reported in Arizona, but Experts are Skeptical

Tipton, Indiana, *Tribune*, April 4, 1935

Phoenix, Ariz., April 4.—Reports that a herd of 100 ibex inhabit the mountains near Kingman have been made to the biological survey at Albuquerque, N. M., by Mohave county officials.

Although representatives of the biological survey declared the reports "impossible," they have agreed to send a field man to the Kingman district to see the strange animals. Ibex are usually found in Asia.

The reports were said to have been filed by J. W. Faulkner, superior court judge, and Sheriff Ernest Graham, who insisted the animals were neither antelope nor mountain sheep.

Boulder Dam Ibex Believed Roosevelt Herd Offspring

Phoenix, Arizona, *Independent Republic*, April 4, 1935

Kingman, Apr. 3.—(AP)—A herd of ibex, whose progenitors are said to have been placed in Nevada by the late President Theodore Roosevelt nearly 25 years ago, is ranging in the Big Bend country above and below Boulder dam.

Elmo Bollinger, Mohave county attorney, and Sheriff Ernest Graham, as well as numerous citizens of this area, aver they have seen the animals, a variety of wild mountain goats, found usually in Asia and Africa.

"While traveling with Sheriff Graham in the mountains I saw six ibex in one group and shortly afterward four more," Bollinger said. "I understand their progenitors were a gift to President Roosevelt and he had them sent to Nevada and the Grand Canyon area."

Engineer Saw Ibex

H. L. Lyon, resident engineer on the Boulder dam road, also reported seeing them.

Government officials, expressing skepticism over the existence of the ibex, have ordered a field man here to examine the animals.

Al Jagerson, service station operator at Willow beach, stated having seen them many times. He estimates 150 in the herd.

"I have seen them swim back and forth across the Colorado river below Boulder dam," he declared. "There have been as many as 40 in one day browsing under the shade trees during the summer near the river adjacent to Willow Reach."

Larger than Goats

No one here seems to remember how many of the animals were released by the former president, noted for his big-game hunting.

Bollinger asserted they have all the characteristics of the ibex. "They are different from the mountain sheep and goats of this section," he added. "They are larger and more powerful than the goats; and have the uniform color of their species, various degrees of brown, usually lighter on the throat, stomach and between the legs than elsewhere."

E. B. Morrison of Kingman, familiar with the habits of the ibex, said the Boulder dam region affords them natural surrounding.

He pointed to the cold winters and the fact the ibex remain upon the crags as near as possible to the snow line, descending the mountains to find uncovered pasturage.

Big Herd Seen

"They usually go about in small bands," Bollinger said, "but sometimes gather into a herd of 100 or more."

"Scores of citizens, in Kingman and vicinity have seen these animals during the last 10 years or more and we believe they are ibex," he concluded.

From Washington came word however that biological survey officials were equally certain the purported ibex were only "mountain sheep."

Scientists also said they "could not have been ibex because that animal is not native to America."

Wild Ibex Reported in Big Bend

Lockhart, Texas, *Post-Register*, May 30, 1935

A herd of wild ibex, a species of mountain goat usually found only in Asia and Africa, has been reported ranging in the Big Bend country.

Numerous citizens of that area insist they have seen the animals though game authorities are openly skeptical.

The theory advanced is that they are descendants of a herd of ibex placed 25 years ago in a game preserve in Nevada by President Theodore Roosevelt.

So-Called Ibex Seen in Arizona May Be a Cross-Breed Species

Oshkosh, Wisconsin, *Daily Northwestern*, July 4, 1935

Flagstaff, Ariz.—(U.P.)—Fastnesses of the little-explored and almost inaccessible Sycamore canyon north-west of here may harbor a species of ibex, hitherto unknown outside Africa, reports reaching here indicated.

Hunting parties, on trips into the wild country, have sighted strange animals, unlike anything known to be native to this section of the United States, they reported.

In each instance, they described the animals' horns as exactly like those of ibex they had seen in zoos or in the trophy collection of Warren P. Lawson, Arizona rancher, who has hunted big game on four continents.

Interest in the strange animals was raised to the point recently that federal authorities were notified and were understood to be sending biological survey experts into the region to attempt to identify the animals.

Lawson, however, said he believed the so-called "ibex" may be the result of strange interbreeding between domestic sheep and mountain goats. He pointed to the fanciful colors of the wool from sheep raised by Navajo tribesmen in support of his contention. The Navajo's sheep, he said, had been inbred and crossbred with goats for so long that strange hybrids were produced.

The rancher-big-game hunter demanded that the state game department protect the "ibex." Interest in them was so great since their discovery, he said, that it was likely hunters would exterminate them.

Mountain Sheep Charges Dropped on Quaint Plea

Reno, Nevada, *Evening Gazette*, April 12, 1938

Oatman, Ariz., April 12 (AP)—Court records revealed today that a jury believed sufficiently in the presence of ibex, a species of old world goat, in Mohave county that three men accused of killing mountain sheep were acquitted.

The jury said the state did not prove conclusively the animals were not ibex, although two deputy state game wardens and a University of Arizona professor testified there was no such animal in Arizona.

Former President Theodore Roosevelt reportedly sent half a dozen ibex to the vicinity of Boulder dam. The U. S. biological survey, however, has never found any record of the ibex being released in the United States.

One defense witness, Jerry Hoffman, who said he had been a miner and prospector in the area between Kingman and Boulder dam thirty years, testified he had seen one hundred of the strange looking animals.

Have You Got Any Ibexes You Want Seen?

Albuquerque, New Mexico, *Journal*, May 10, 1938

Flagstaff, Ariz., May 9 (AP)—An Arizona ibex is worth $100.

Such a reward was offered by T. E. McCullough, president of the Arizona Game Protective Association, to anyone who could locate a live, uncaged ibex in the state.

There has been much controversy over the existence of ibex in Mohave County.

Several men in Mohave County charged with killing mountain sheep were released when they claimed the animal slain was an ibex, descendant of a herd released in Northern Arizona by Theodore Roosevelt about 1900.

"Whatisit" Mystery Bobs Up Again in Alamogordo Area; May Start Search

Gallup, New Mexico, *Independent*, February 28, 1940

Alamogordo, Feb. 28 (AP)—A 40-year-old argument over the "whatisit" of the San Andreas mountains has cropped up again—this time with demands for an expedition to hunt one of the strange animals.

"It's an ibex." declare the ranchers in the little known mountains that lie between two southern New Mexico deserts—the White Sands and Jornada del Muerto—journey of death.

Scoffs State Game Warden Elliott Barker:

"There are no ibex on the American continent; probably a mountain sheep."

"But," swear the old time stockmen, "this animal isn't a mountain sheep or a mountain goat. It has hair, not wool, and straight horns."

The argument has been going that way ever since the late Bob Burck killed a strange animal in the upper reaches of the San Andreas. The old timers insist it resembled pictures of an ibex and that experts at the New Mexico Agricultural College so identified it.

Conclusive proof of the story, they maintain, is "Ibex corral" so called because Burck nailed up the head there.

A year or more ago, the game department sent an expert into the San Andreas and found nothing—the ranchers say he spent only a few hours and didn't reach the "whatisit's" range.

To the proposal of an expedition into the inaccessible heights, Warden Barker has been adamant, warning them, the ranchers say, that if a "whatisit" were bagged and it turned out to be a protected animal, the hunter would be prosecuted.

M. C. Cauthen, head of the Otero County Game Protective Association, entered the negotiations recently in an effort to bring the two factions together. Frank Andregg, one of the San Andreas ranchers, added an offer to guide a "whatisit" hunt without charge.

The Roosevelt Introduction Theory

One of the interesting aspects of cryptozoology is that because all mystery animals are folkloric (whether or not they exist in actual fact, or are undescribed or unrecognized species), there is a tendency to attract other folkloric elements. In extreme cases, this leads to mystical elements (i.e., paranormalism) being attached consciously or unconsciously by witnesses or investigators. But, here we have another layer of folklore, the authoritarian origin story. There are stories all across North America of wildlife officers furtively reintroducing predators (eastern cougar, coyotes, rattlesnakes) in order to control prey populations. "Somebody" saw them do it, so it must be true. (This isn't just North American; similar stories are found in Europe, and not that long ago we saw folklore coming from the Middle East about "strange animals" alleged to be released by the U.S. military to attack livestock.) The idea that Theodore Roosevelt released ibex into the southwest is clearly folkloric, and as of yet, I've seen no evidence that it is true. It probably spread because it sounds like the sort of thing that conservationist Roosevelt might have attempted. One journalist (Halloran 1970) suggested that the Roosevelt story was based on chamois that were promised to Roosevelt by the Swiss government to release in Yellowstone, but didn't arrive until Roosevelt was out of office, at which point they were just turned over to the National Zoo. In this case, the story is more innocuous than "predator release" tales, but the narrative is very similar.

Ibexes in North America Today

While there may not be a native species of American ibex, we now have free-roaming introduced ibexes in New Mexico. Introductions began in the 1960s, as part of a state program to release captive-born offspring of several ungulates for big game hunting. A brief summary of the more significant ungulate activities:

1950: New Mexico begins introducing exotic ungulates, first with 52 aodad into Canadian River Canyon (Murphy 1969).

1964: New Mexico placed Iranian ibex, Siberian ibex, and gemsbok (Oryx) in the Red Rock (Grant County) state-operated game pastures (Murphy 1969). Once the animals multiplied, offspring would be released into the wild.

1969: Oryx were released into the White Sands Missile Range, New Mexico (Anon. 1970).

1970: New Mexico Game and Fish released 15 Iranian ibex into the Florida Mountains of southwest New Mexico (Anon. 1970).

1973: By 1973, there were an estimated 40-50 Iranian ibex in the Florida Mountains, and an additional 10 were introduced to the herd (Anon. 1973). Another six were added later in the year (Anon. 1973b).

1975: New Mexico held their first ibex hunt, allowing four hunters to acquire permits (Anon. 1975).

1975: New Mexico had to nix its plan to introduce Siberian ibex onto federally-owned Ladron Mountain when the Bureau of Land Management refused to allow it, citing new changes in the National Environmental Policy Act (Anon. 1975b).

1976: New Mexico began transferring the sixty-some Siberian ibex from their Red Rock facility to Canadian River Canyon (Archibald 1976). This introduction apparently failed (Mungall 2007), dying out in the 1980s (Trennel 2010), though there may be occasional rare sightings.

1977: Outdoor editor Bill Quimby (1977) noted that wanderers from the New Mexico exotic introductions are occasionally reported from the Arizona side of the border, but weren't at that time taking root.

1985: An aerial survey of Iranian ibex in the Florida Mountains counted 695 in three hours time (Taylor 1985).

Today, several species (and hybrids) of ibex can also be found roaming private game ranches in various southwestern states. Escapes do happen, though game officers react quickly, so as to make certain exotics won't spread diseases or outcompete native species.

Conclusion

Obviously, misidentifications are the likely explanation for most, if not all, alleged American ibex sightings. Certainly, sighting reports should always be treated cautiously. Striving for the best evidence possible requires physical material for examination and testing. If I were to suggest any location as a possible starting point for further research, it would be in Alaska. After all, if there were a species of ibex living in the mountainous areas of southern Alaska, it is possible that an Alaskan hunter may already have possession of a skull and horns, and not recognize its scientific significance (not an uncommon occurrence when it comes to undescribed species). This direction might be worth some effort.

The obvious rejoinder is, why bother investigating at all? Isn't this a clear open-and-shut case of mistaken identity? Interestingly enough, there is one further bit of circumstantial evidence: ibex-like representations from a number of Native American petroglyphs.

In one series of petroglyphs in Havu Supai, archaeologists discovered the outlines of what are clearly ibex-like figures. The horns appear too straight (while recurved) and long to be obviously representative of mountain goats or bighorn sheep. Similar petroglyphs are found elsewhere in various western states. There is more than one possible explanation, of course:

1) The representations are exaggerated, poorly carved, or show unrealistic perspectives, and actually represent recognized species (likely bighorn sheep).

2) The carvers identified bighorn sheep with aberrant horns as separate from the typical form, and carving them distinctively.

3) The carvings do in fact represent a true American ibex, which may or may not still exist.

Hava Supai Canyon, Arizona (Hubbard 1925)

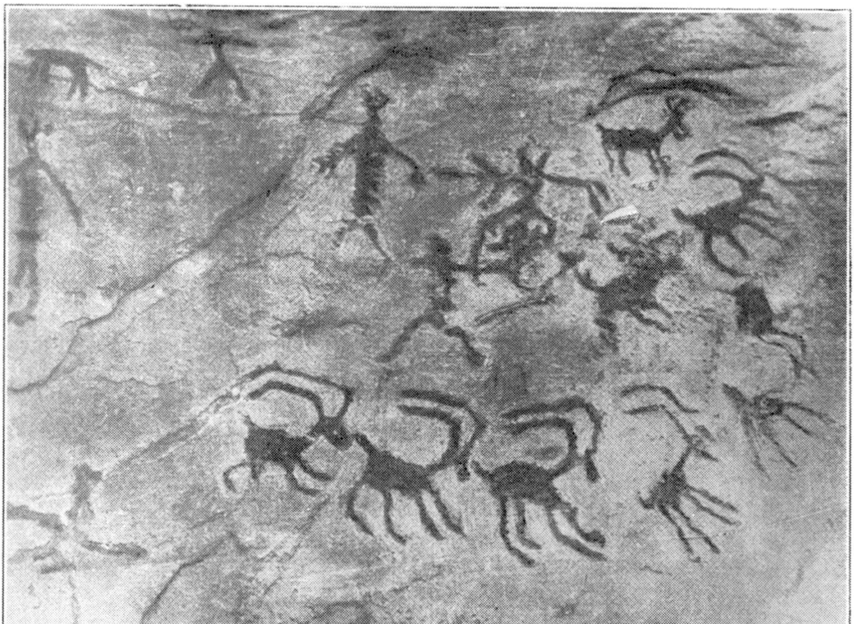

Lee Canyon, Arizona (Hubbard 1925)

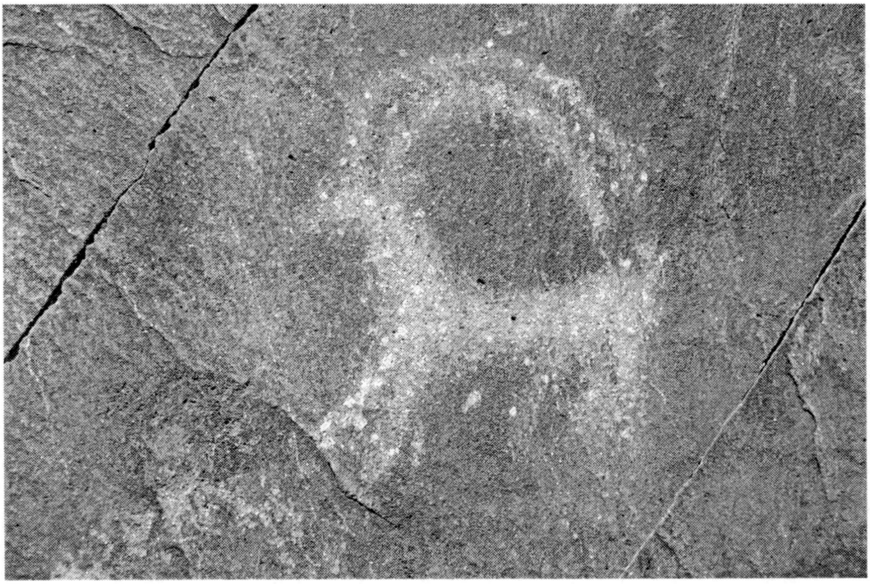

An Actual Ibex Petroglyph, Kazakhstan (Denis Sukhov)

Petroglyphs of Rocky Mountain Goats,
near Irvins, Utah (Curtis Bates)

BIGHORN SHEEP PETROGLYPHS,
WANAPUM RECREATIONAL AREA, WASHINGTON
(EUGENE KALENKOVICH)

A UTAH PETROGLYPH (RALF BROSKVAR)

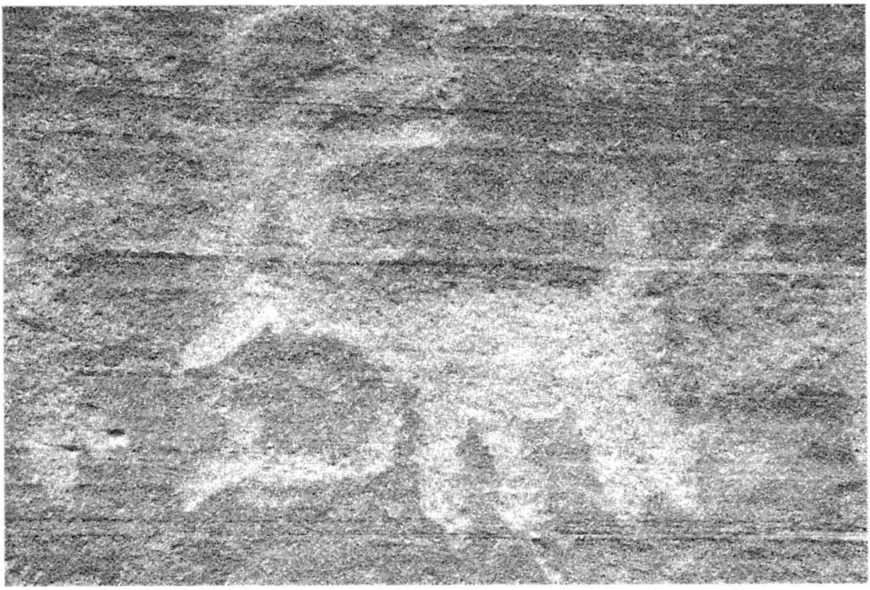

VALLEY OF FIRE STATE PARK, NEVADA (VLADISLAV GURFINKEL)

At this point, there is no fossil evidence that ibex crossed the Bering Strait, though a number of other ungulates did. Unless confirmative physical evidence is ever brought forward, questions regarding the American ibex's true identity will remain ambiguous at best.

References

Anonymous. 1912. An American ibex. *Outing* 60(2): 249-250.

Anonymous. 1970. Game and Fish Department releases 15 Iranian ibex. Las Vegas, New Mexico, *Optic* (December 9): 4.

Anonymous. 1973. More ibex released in Florida Mountains. Las Cruces, New Mexico, *Sun-News* (February 22): 16.

Anonymous. 1973b. To expand ibex herd. Hobbs, New Mexico, Daily News-Sun (October 2): 8.

Anonymous. 1975. Deming hunter gets his ibex. Deming, New Mexico, *Headlight* (January 16): 1.

Anonymous. 1975b. State exotic game program may end. Silver City, New Mexico, *Daily Press* (July 29): 2.

Archibald, Bill. 1976. Game Department transfers ibex. Silver City, New Mexico, *Daily Press* (February 18): 2B.

Arment, Chad. 2011. The curious creature. *Flying Snake: A Journal of Cryptozoology, Folklore and Forteana* 1(1): 5-6.

Cane, Claude. 1903. *Summer and Fall in Western Alaska: The Record of a Trip to Cook's Inlet after Big Game*. London: Horace Cox.

Cohen, Claudine. 2002. *The Fate of the Mammoth: Fossils, Myth, and History*. trans. William Rodarmor. Chicago: University of Chicago Press.

Dall, Wm. H. 1895. The St. Elias bear. *Science* 2(30: July 26): 87.

Grant, Madison. 1904. The rocky mountain goat. *Annual Report of the New York Zoological Society* 16: 231-261.

Halloran, Art. 1970. American ibex is just a myth. Lawton, Oklahoma, *Constitution–Morning Press* (May 3): 8C.

Hornaday, W. T. 1915. The irrepressible American 'ibex.' *Zoological Society Bulletin*. 2(24): 327-328.

Hornaday, William T. 1919. *The American Natural History*. New York: Charles Scribner's Sons.

Hubbard, Samuel. 1925. *The Doheny Scientific Expedition to the Hava Supai Canyon, Northern Arizona*. San Francisco: Sunset Press.

Kurtén, Björn, and Elaine Anderson. 1980. *Pleistocene Mammals of North America*. New York: Columbia University Press.

McGuire, John A. 1921. *In the Alaska-Yukon Game-Lands*. Cincinnati, OH: Stewart Kidd.

Mungall, Elizabeth Cary. 2007. *Exotic Animal Field Guide: Non-native Hoofed Mammals in the United States*. College Station, TX: TAMU Press.

Murphy, Andy. 1969. African game in New Mexico: Experiment in ecology. . . Silver City, New Mexico, *Daily Press* (February 22): 12.

Osgood, Wilfred H. 1909. The big game of Alaska. *National Geographic* 20(7): 624-636.

Quimby, Bill. 1977. Wayward sheep find Arizona. Tucson, Arizona, *Citizen* (July 22): 3F.

Radclyffe, C. R. E. 1904. *Big Game Shooting in Alaska*. London: Rowland Ward.

Sheldon, Charles. 1919. *The Wilderness of the Upper Yukon*. Second edition. New York: Charles Scribner's Sons.

Shields, G. O. 1902. Is there an ibex in this country? *Recreation* 16(2): 114.

Shields, G. O. 1904. Photographing in the Canadian Rockies. *Recreation* 20(2): 89-98.

Taylor, Mark. 1985. The exotic ibex is a prized target. Hutchinson, Kansas, *News* (April 21): 36.

Trennel, Tony. 2010. Possible unusual sightings around White Pine, CO. *New Mexico Shooting News* (April-September): 4.

Williams, Milt. 1977. Early hunters went on wild ibex chase. Pocatello, Idaho, *State Journal* (April 3): B5.

The Popcorn Fish

Chad Arment

Not all mystery animals are large and monstrous. Kevin Stewart, a Canadian correspondent, brought this particular cryptid to my attention several years ago. It is a great example of a mystery animal that has managed to escape the notice of the general public—and even that of the cryptozoological community.

The Popcorn Fish was first reported in the summer of 1953 (Anon. 1980). The cook for a geological survey party in the Yukon caught a strange fish ("distinguished by having a large number of bumps on its head, which resembled popcorn") in a lake 500 km north (slightly northeast, actually) of Whitehorse. The cook, Al Martin, had never seen anything like it, and he named it the Popcorn Fish. Then, he fried it up for a meal that he shared with two others.

Dr. Cas Lindsey, of the University of British Columbia, a freshwater fish specialist, heard this story, and the stories of bush pilots and others in that part of the Yukon. With Dr. Don McPhail, he set out in 1960 to the lake in question in an attempt to net the fish. Unfortunately, they only caught suckers, sculpin, and grayling. The following two stories were reported in regional papers regarding this expedition.

Tracking Down 'Popcorn' Fish

Fairbanks, Alaska, *Daily News-Miner*, June 22, 1960

Vancouver, B.C., June 22, (AP)—University of British Columbia scientists will seek the mysterious "popcorn" fish, said to be in the Yukon's Peel River headwaters, during this season's collecting expeditions,

Under the direction of Dr. C. C. Lindsay, the expeditions will go into remote areas of northern B.C., Yukon, Alaska and Bering Sea islands.

The "popcorn" fish has been included consistently in reports from bush pilots, who say the fish has head lumps resembling popcorn.

This year's collections for the Institute of Fisheries already has yielded two specimens previously unknown in B.C. waters—the nine-spined stickleback and the emerald shiner, both found at Fort Nelson.

Scientists Fish in Alaska For Clues to Ancient Era

Fairbanks, Alaska, *Daily News-Miner*, October 28, 1960

Vancouver, B.C., (Special)—Belief that a land bridge once connected Asia with North America has been strengthened as a result of a summer expedition by scientists from the University of British Columbia's Institute of Fisheries.

The expedition, headed by Dr. C. C. Lindsey, found fresh water fishes on St. Lawrence Island in the Bering Sea between Alaska and Russia. Dr. Lindsey, curator of fishes at UBC claims the three species of fresh water fishes found could not have reached the island except through the fresh water channels of a land bridge.

Three Specimens

The three specimens found on the island, which is about 100 miles in length and within sight of the Russian coastline, are the Alaska black fish, the Arctic grayling and the slimy sculpin.

The Alaska black fish is found only in the Arctic near the Bering Sea. The grayling is a sports fish.

The summer expedition sponsored by an annual grant from Dr. H. R. MacMillan and the Arctic Institute, yielded

> 1300 pounds of fish which was preserved for study by scientists at UBC.
>
> Dr. Lindsey and his associates also made a trip to the headwaters of the Peel River in the Yukon in search of the "popcorn" fish, so-called because of numerous bumps on its head. Despite repeated attempts to net the fish the party was unable to capture one.
>
> Indian tales of fresh water flying fish were investigated at another lake reached by chartered plane.
>
> The flying fish proved to be dwarf Arctic grayling, a small fish with large fins. The lake investigated was so overpopulated with the fish that their constant leaping into the air for food sounded like rain falling on the lake surface.

This was apparently the only official expedition in search of the strange fish. Dr. Lindsey noted that it was possible that the original Popcorn Fish was an ordinary sculpin or sucker "with some type of abnormal head growths, possibly caused by parasites" (Anon. 1980). He did not rule out the possibility, however, that it could be a unique species surviving in this remote area, stating, "If Canadians are prepared to believe in the Sasquatch and in Ogopogo, they should keep an open mind about more plausible creatures like the popcorn fish."

In any case, this strange mystery fish has marked its spot on the map. After reviewing Dr. Lindsey's notes, the Canadian Permanent Committee on Geographical Place Names officially recognized Popcornfish Lake. The lake can be found at 65.4639°N, 133.8056°W.

References

Anonymous. 1980. On the trail of the elusive popcorn fish. *UBC Reports* (University of British Columbia) 26(16: August 27): 2.

The Hungarian Reed Wolf (*Canis spp.*): Speculations on an Extirpated and Forgotten Population

Tomasz Pietrzak and Miklós Heltai

Introduction

The Hungarian reed wolf was a small wolf-like or large jackal-like canid. It was historically classified as an extinct subspecies of the grey wolf named *Canis lupus minor*, but its phylogenetic origin wasn't well explained. The reed wolf was first discussed by zoologist Kramer in 1756. Ogérien (1863) described a small Austro-Hungarian wolf as *Canis lupus minor*, a subspecies of the grey wolf that lived in western Hungarian territories. (Zoologists now believe that *C. l. minor* was the same as the common European wolf, *Canis l. lupus*.) Discussion and debate on the taxonomy of the Austro-Hungarian reed wolf was undertaken by later European scientists and naturalists. Mojsisovics (1897) first suggested the reed wolf was the diminutive *Canis lupus minor*, and this was for many years generally accepted.

Various studies in the twentieth century were carried out regarding the taxonomy of the reed wolf. Dr. Gyula Éhik (1938) reclassified the reed wolf, stating it was the same canid as *Canis aureus hungaricus*. According to many researchers, *C. a. hungaricus* is morphologically the same as the Dalmatian jackal. Kretzoi (1947) proposed that the reed wolf was a distinct Hungarian jackal, giving it a new scientific name: *Canis aureus ecsedensis*. Dr. Eugen Nagy (1956) adhered to this classification and supported the historical extirpation of the unique reed wolf. János Szunyoghy (1957) announced the invalidity of *hungaricus* as a separate subspecies. He also proposed (Szunyoghy 1959) that the reed wolf was simply a larger form of European golden jackal (not even a separate subspecies of *Canis aureus*). Other scientists besides Szunyoghy and Nagy have also speculated that it wasn't a variety of wolf, but just a golden jackal or possibly even a feral domestic dog. In the past, after the Turkish wars, feral domestic dogs bred outside village areas. It is possible that many dogs remain in the wild. However the theory that these were responsible for reed wolf reports has not stood scientists' scrutiny. Tratz (1958) conducted detailed studies on the few known museum specimens in an attempt to clarify the reed wolf's identity, and his proposal that it was a morphologically small wolf was well-received (Shuker 2007).

One interesting idea without confirmation in later years, was that the once-mighty flood plains of Hungary influenced the development of smaller, short-haired wolves, named later the reed wolf (Bioport 2008). As previously explained, many scientists have investigated the viability of the subspecies theory and even whether it was a distinct species of European wolf. After all, if the canid was a new species or a morphological variant, it would be a windfall for many research institutes, and in Hungary and Austria it could have been an important tourist attraction.

Bauer (1960) ended the debate by eliminating the reed wolf as a subspecies of the grey wolf from the list of Austrian fauna. It was argued that the data about this animal are too poorly documented and it was nothing more than the common golden jackal—now the officially accepted view. There have been other attempts to classify the reed wolf as a local jackal subspecies, a jackal with genetic introgression from wolf genes, or an inbred jackal population endemic to Central Europe.

This animal lived on the moors along the border of Austria and Hungary. G. H. Kramer, in his first book (1756) of Austrian fauna, mentioned large numbers of reed wolves (in German, *rohrwolf*) living in the region of Lake Neusiedler See, where the golden jackal has re-settled again in the early twenty-first century (Arnold et al. 2011). Nowadays, the animal is considered extinct and its taxonomy can be determined only on the basis of museum specimens located primarily in natural history museums in Hungary. (See more in the work of Prof. Éhik. His original material, e.g. the voucher specimen of *Canis aureus hungaricus*, is in the Hungarian Natural History Museum.) Thanks to Ludwig Heinrich Jeitteles we can also observe a few specimens of this wolf-like canid at the Institute of Natural History Museum in Vienna.

DISCUSSION ON THE GOLDEN JACKAL

The reed wolf first appeared as a European cryptid in a famous book by cryptozoologist Dr. Karl P. N. Shuker, *Extraordinary Animals Worldwide*, in 1991. The question of its phylogeny remained: what animal could be the true reed wolf? Was it a jackal with wolf-like characteristics, a wolf that looked like a jackal, or something else? One hypothesis even suggested that the name *reed wolf* was given to both the grey wolf and the golden jackal, as a generic name for any Hungarian canid, recognized or unrecognized. The reed wolf became extinct due to excessive hunting and destruction of its environment, and as we know it was endemic, such actions would easily have destroyed such a small population. Apparently, the "diminutive wolf" population inhabited open woods and areas overgrown with shrubs and reeds (hence the local name). Some people speculated that it was a unique local hybrid between the golden jackal and the grey wolf, which went extinct as a result of destructive human activity. We can't explicitly confirm any hypothesis without genotyping the peculiar specimens. Perhaps in the future scientists will confirm or reject these theories through the analysis of mtDNA or nuclear microsatellites by PCR method.

Has the native jackal survived? It is not certain, but it seems that it hasn't, because in this case the golden jackal could not take advantage of the same niche as well as its larger cousin, and also its greatest enemy, the grey wolf. A native jackal would likely be absorbed into the invading jackal population from the south. Nowadays, of course, there are wolves in these areas, but they do not form a stable population and belong to the typical subspecies. Wolves in Austria appear sporadically, and the population is not larger than 20 individuals. In Hungary, the population consists of one known breeding pack in the northern part of the country, occupying a vast territory that includes both Slovakia and Hungary (Hausknecht et al. 2009).

Currently the reed wolf habitat in Hungary and Austria is inhabited by regenerated populations of the golden jackal as well as by occasional European grey wolves (in the case of Hungary see Bihari et al. 2007; Heltai 2010). These animals took up the niche occupied by the former extirpated canids. Wolves prefer deciduous mixed forests, while jackals prefer open steppe and extended marshy areas. On the other hand, jackals try to avoid areas with lots of snow. According to Kryštufek et al. (1997), the jackal is rare or absent from the hilly regions of its range here. The golden jackal originated in India, from where it expanded into northern, eastern and western Africa, and into southeastern Europe, including Greece, Turkey, Serbia, Hungary, and Slovakia. The species is now spread throughout the Balkans. The past twenty years have led to an increased population and expanded range of *Canis aureus moreoticus* in these lands that it formerly inhabited (see more in Szabo et al. 2009; Arnold et al. 2011).

The golden jackal (*Canis aureus*), the indigenous predator of Hungary, is listed in the Hungarian Red Data Book as an extinct species. The original population of native golden jackals in Hungary declined dramatically at the beginning of the second half of the 20th century and was finally extirpated due to habitat fragmentation and hunting pressure (Kryštufek et al. 1997). The last known specimen of this indigenous population was shot in 1942 in near the village of Derecske, situated in the Hungarian Great Plain. This species appeared again in south-central Europe as an invasive non-native, extended its range northwest, where it arrived in Hungary and Serbia (Szabo et al. 2009; Heltai, unpublished). It was first reported in 2006 from southern Moravia in the Czech Republic (Koubek and Cervený 2007). It has been documented in northeastern Italy (1988), Germany (1996), and Slovakia (1989) (Hell & Bleho 1995).

Golden jackal returned to southern Hungary as early as the late 1970s, with the first breeding pairs being detected near the southern border in Transdanubia, then between the River Danube and

Tisza. Independent data collection and analysis confirmed that golden jackals have settled in Hungary. The population now seems to be quite large in the southeastern part of the country. The rate of expansion and population growth are typical for invasive species. The population in Hungary increased in numbers dramatically after the early 1980s, when immigrating individuals reappeared in its former territories, leading in 1991-92 to the re-establishment of a viable population (Arnold et al. 2011). According to official hunting bag data, the number of bags reported has continuously increased, with 11 jackals shot in 1997 and a total of 163 specimens shot up to 2006 (Szabo *et al.* 2009).

What are the reasons for the rapid spread of the invading golden jackal in Europe? The answer is simple: deforestation and the expansion of steppe habitat and open spaces with swampy areas created environmental conditions that were very favorable for the jackals. Other advantages to jackals included the reduction of the wolf population, which ensured a larger share of easy prey (domestic livestock and carcasses), the abandonment of poisoned baits, and the natural cycles of population dynamics. We also have to mention that the golden jackal now has only one true competitor with European-wide occurrence and a high population—the red fox (*Vulpes vulpes*). The red fox uses the same habitats (Giannatos 2004; Giannatos *et al.* 2005) and food resources (Lanszki *et al.* 2006; Lanszki & Heltai 2010) as the golden jackal, but the golden jackal is stronger than the red fox. The jackal also has a better reproductive strategy and adapts better to different habitats, from the marshland to the dry grassland (Szabó *et al.* 2008).

CONCLUSION

Can we explain the origin of the reed wolf? The reed wolf was probably an indigenous golden jackal, a locally unique "genetic source population," also known as an Evolutionary Significant Unit (ESU), now extinct. Most likely, the last representative of the unique *reed wolf* form of canid was killed March 4, 1950, in the state forests of the Austrian district of Rotmos. In Hungary, the last known specimen was shot in 1942. It was likely a historically separated larger form of the golden jackal, adapted to life in the swamp reeds, with a unique level of genetic diversity and greater differentiation from other populations through reproductive isolation (high FST index with isolation by distance). Mythology likely also played a part, if the golden jackal was sometimes confused with a locally occurring population of European grey wolf, bursting into the legend of the "reed wolf."

Nowadays, Hungary is inhabited by the invasive southern golden jackal, a new opportunistic predator, but it is really not the same as the Hungarian reed wolf, or more correctly the *reed jackal.* However, without DNA samples from the few remaining museum specimens, we will never know for certain the real secret of the "wolf from the reeds."

Acknowledgments

We would like to thank Edyta Swider for the English corrections of the text. We are grateful for Chad Arment for his comments on the text.

Literature References

Arnold, J., A. Humer, M. Heltai, D. Murariu, N. Spassov, and K. Hackländer. 2011. Current status and distribution of golden jackals (*Canis aureus* L., 1758) in Europe. *Mammal Review* 41: 1365-2907.

Bauer, K. 1960. Die Säugetiere des Neusiedlersee-Gebietes. *Bonn. Zool. Beitr.* 11:141-344.

Bihari, Z., G. Csorba, and M. Heltai. (eds) 2007. *Magyarország emloseinek atlasza* /The Atlas of Hungarian Mammals/. Kossuth Kiadó, Budapest, 360 pp.

Bioport. 2008. *James G. Frazer: Aranyág* (review). http://www.bioport.hu/masvilag/mitoszok/nadi_farkas.htm (Disabled link, now located at Archive.org.)

Éhik, Gy. 1938. Sakál vagy nádifarkas hazánkból. (Jackal or Reedwolf from Hungary). *Annales Musei Nationales Hungarici* 31: 11-15.

Giannatos, G. 2004. Conservation action plan for the golden jackal (*Canis aureus* L. 1758) in Greece. WWF Greece, Athens, 1-47.

Giannatos, G., Y. Marinos, P. Maragou, and G. Catsadorakis. 2005. The status of the golden jackal (*Canis aureaus* L.) in Greece. *Belgian Journal of Zoology* 135: 145-149.

Hausknecht, R., Á. Szabó, G. Firmánszky, R. Gula, and R. Kuehn. 2010. Confirmation of wolf residence in Northern Hungary by field and genetic monitoring. *Mammalian Biology* 75: 348–352.

Hell, P., and Š. Bleho. 1995. Novodobý výskyt šakala obycajného (*Canis aureus*) na Slovensku. /Contemporary occurrence of jackal (*Canis aureus*) in Slovakia./ *Folia Venat.* 25: 183–187.

Heltai, M. (ed.) 2010. *Emlos Ragadozók Magyarországon* /Mammal predators in Hungary/. Mezogazda Kiadó, Budapest.

Kerschner, T. 1959. Der "Rohrwolf" von Erhagen bei Enns. *Jahrbuch des Österr. Arbeitskreises für Waldtierforsch* 1959: 40-46.

Koubek P., and J. Cervený. 2007. The golden jackal (*Canis aureus*)—a new mammal species in the Czech Republic. *Lynx* 38: 103-106.

Kramer, G. H. 1756. *Elenchus Vegetabilium et Animalium per Avstriam Inferiorem Observatorum*. Vienna, Prague, Tergesti: Th. Trattner.

Kretzoi, M. 1947. New names for mammals. *Ann. Mus. Nat. Hung.* 40: 285-287.

Kryštufek, B., D. Murariu, and C. Kurton. 1997. Present distribution of the golden jackal *Canis aureus* in the Balkans and adjacent regions. *Mammal Review* 27: 109-114.

Lanszki, J., Heltai, M., and Szabó, L. 2006. Feeding habits and trophic niche overlap between sympatric golden jackal (*Canis aureus*) and red fox (*Vulpes vulpes*) in the Pannonian ecoregion (Hungary). *Canadian Journal of Zoology* 84(11): 1647-1656.

Lanszki, J., and M. Heltai. 2010. Food preferences of golden jackals and sympatric red foxes in European temperate climate agricultural area (Hungary). *Mammalia* 74: 267-273.

Mojsisovics, A. 1897. *Das Tierleben der österreichisch-ungarischen Tiefebenen*. Alfred-Hölder: Wien.

Nagy, J. 1956. Das ausgerottete ungarische Rohrwolf (*Canis lupus*) war kein Schakal (*Canis aureus*). *Säugetierkundliche Mitteilungen* 4 (4): 165-167.

Ogérien, F. 1863. *Histoire Naturelle du Jura et des Départements Voisins*. Tome III, Zoologie vivante. éd Masson and Fils, Paris. A. Robert et Gauthiers Fils. Lons-le-Saunier.

Shuker, K. P. N. 2007. *Extraordinary Animals Revisited: From Singing Dogs To Serpent Kings*. Bideford: CFZ Press.

Szabó, L., M. Heltai, and J. Lanszki. 2008. Factors that influence the Golden jackal's (*Canis aureus* L. 1758) spreading in Hungary. 92nd Annual Meeting of the German Society of Mammalogy. Vienna, 14 to 18 September 2008. Abstracts of Oral Communications and Poster Presentations. *Mammalian Biology* 73: 41.

Szabó L., Heltai M., Szucs E., Lanszki J., Lehoczki R. 2009. Expansion range of the golden jackal in Hungary between 1997 and 2006. *Mammalia* 73(4): 307-311.

Szunyoghy J. 1957. Systematische revision der ungarlandischer schakals, gleichzeitig eine bemerkung über das rohrwolf-problem. *Annales Historico-Naturales Musei Nationalis Hungarici* 78: 425-433.

Szunyoghy, J. 1959. Bírálat Nagy Jen "Berichtigung zur Szunyoghys Systematischer Revision des ungarländischen Schakals, gleichzeitig eine Bemerkung über das Rohrwolf-Problem" cím dolgozatáról. *Vertebrata Hungarica* 1(2): 209-214.

Tratz, E. 1958. Ein betrag zum kapitel 'Rohrwolf' *Canis lupus minor* Mojsisovics, 1887. *Säugetierkundliche Mitteilungen* 6: 160-162.

Freshwater Seals in Alaska and Canada

Chad Arment

Although uncommon, there are several populations of seals around the world that are resident year-round in freshwater lakes. The Lake Baikal seal (*Phoca sibirica*) of Siberia is best known, and there are ringed seal subspecies endemic to lakes in Finland (*Phoca hispida saimensis* of Lake Saimaa) and Russia (*Phoca hispida ladogensis* of Lake Ladoga). Additionally, there is a seal (*Pusa caspica*) endemic to the brackish Caspian Sea. The Caspian Sea has a salinity approximately one third of typical seawater.

In North America, there are two recognized populations of harbor seals (*Phoca vitulina*) that live in freshwater lakes. One is in Iliamna Lake, Alaska, and the other is found in Lower Seal Lake of the Ungava peninsula of northern Quebec. This Ungava seal has been designated a distinct subspecies, *P. v. mellonae*, though some biologists have suggested that this population is composed of wandering seals from marine waters that migrate occasionally over land (Reijnders et al 1993). Because the populations in Iliamna Lake and Quebec are fairly small, and are of little economic importance (beyond some local subsistence hunting), research on them has been sporadic. There is still quite a bit that has yet to be uncovered about their biology. There are a few other questions regarding freshwater seals on this continent, also. Sometimes seals are reported from freshwater lakes or rivers where no populations are known. And, there is the mystery of the dwarf seals, reported from a lake in the Canadian Arctic Islands.

For most out-of-place seal mysteries here, we are dealing with the harbor seal, *Phoca vitulina*, which is so well-known for traveling up waterways that it is sometimes called the ranger seal. On the east coast, it was historically known to wander into Lake Champlain and Lake Ontario (Miller 1899). On the west coast, it sometimes follows a river as far inland as 300 kilometers (Baird 2001).

Iliamna Lake Seals

A population of Pacific harbor seals (*Phoca vitulina richardii*) is found in Lake Iliamna of southwest Alaska. Subsistence hunters catch seals here year-round (Fall et al 2006). A survey in the summer of 2008 noted that at least 235 seals were found in the lake. Local hunters say these seals are "fat and healthy" throughout the year, unlike the marine seals. This is probably due to a healthy diet of mostly salmonid fishes (Hauser et al 2008).

The biggest question has been whether all of these seals are year-round residents. Lake Iliamna is connected to Bristol Bay by the 75-mile-long Kvichak River. While some suggested that the seals retreat to the Bay during the winter, Withrow and Yano (2009) noted that "there are no known accounts of immigration or emigration." In 2008, a photographer, Scott Dickerson, posted a photo on his blog showing seals on Iliamna Lake taken on March 9 of that year. An ecotourism facility on the lake, Rainbow Point Lodge, noted on their website (2011) that they went on a "midwinter expedition" and successfully located seals. However, Withrow et al (2011) ran six aerial surveys from April to November in 2010, and concluded that while "seals do pup in the lake and some seals obviously over-winter, ... it seems unlikely that most seals do." While more than 200 seals had been found during the aerial surveys in late spring and summer, they only found a handful of seals in late fall and early spring.

Ungava Seal

Following up rumors that seals lived in the inland lakes of Quebec's Ungava peninsula, a 1938 expedition sponsored by the

Carnegie Museum set out to investigate these stories and explore the region (Twomey and Herrick 1942). The trip was instigated by a sealskin handy-bag noticed by scientist J. K. Doutt while he was working on the Great Whale River in 1935. It appeared to be different from the common harbor seal, and the Native Canadian who owned it said it was made from a *kasagea*, from the Seal Lakes in the interior of the Ungava peninsula, "almost two hundred miles inland and fully eight hundred feet above sea level." (The instigating bag turned out to be a red herring, being from a harp seal (Doutt 1942).) Prior to this point, while biologists had accepted that seals might be present in the Seal Lakes (Doutt 1942), they had no idea that a seal distinct from the marine population of harbor seals might be responsible for sightings. Doutt and Twomey went on the 1938 expedition, and after numerous difficulties, finally managed to obtain a specimen. Eventually, Doutt (1942) described the Ungava seal as a subspecies of the harbor seal, *Phoca vitulina mellonae*. The recognized population lives in Lower Seal Lake of the Lacs de Loups Marins. There is native testimony that the seal was hunted, historically, in Upper Seal Lake, but it does not appear to be there now. The name Lacs de Loups Marins (Seal Lakes) itself goes back to French cartographer Nicolas Bellin in 1744 (COSEWIC 2007).

Regarding the ethnoknown name given for the Ungava seal, Twomey doesn't state the exact provenance, but apparently it is an Inuit term, rendered *qasigiaq* in the COSEWIC (2007) report. The latter also notes that the Cree call the Ungava seal *nuchimu-achikw* or *achikunipi*. (Doutt (1954) gives the Cree name as "At-chook" or "At-chuk.") The seal is described by the Cree as being smaller and darker than saltwater harbor seals, with different behaviors and a different taste (COSEWIC 2007). Fur-traders have traditionally noted that the freshwater seal skins are of finer quality than that of the marine seals.

Specimens are few, and there have only been a few detailed accounts of these seals in the scientific literature. Interestingly, one Ontario youth canoe-trip camp, Keewaydin Temagami, has visited the lake to look for seals on two occasions, during the early summers of 2000 and 2005. The 2005 trip located several seals (Chapin 2005).

The designation of subspecies for this seal is still under contention. It was primarily differentiated due to "an unusually dark pelage and an enlarged coronoid process on the mandible ... and on the presumption that the population had been isolated for approximately 5,000 years, trapped by the Ungava peninsula's isostatic rebound since the last ice age" (Smith, Lavigne, and Leonard 1994). Some biologists argue that the Ungava seal is not distinctive enough from marine populations of the harbor seal, and that it is likely just a population originated from lost seals that continues to be supplemented by occasional marine wanderers. A recent review of the subspecies (COSEWIC 2007) notes that "the balance of evidence indicates long-term, year-round residency in the Lacs des Loups Marins area." Ungava seals pup much earlier than do marine harbor seals (COSEWIC 2007), which may indicate reproductive isolation. Certainly, a different reproductive cycle would make it very difficult for potential wanderers from the ocean to easily incorporate into the freshwater population.

Accurate population count is difficult to estimate, though probably is in the 100-600 individuals range, with fewer than 250 being mature (Reijnders et al 1993; COSEWIC 2007). The region's Cree and Inuit note the seal can be found in "Lacs des Loups Marins, Petit Lac des Loups Marins, and Lac Bourdel, with some reports of animals having once been in Lac à l'Eau Claire" (COSEWIC 2007; Reijnders et al 1993). A wider range has been suggested by interviews with Inuit hunters by Hydro-Québec contractors, with sightings or killed seals from "Lac Guillaume-Delisle, Rivière Nastapoca, Rivière Boniface, Rivière Niagurnaq, Rivière Kuunga, Rivière Longland, Lac Tasialuk, and Lacs des Loups Marins" (COSEWIC 2007). The advent of rifle hunting apparently resulted in a reduced range for these seals, particularly in areas easily accessible by local Cree and Inuit hunters (COSEWIC 2007).

Mansfield (1967) noted that Inuit traditions noted a wider range for the seal in the Ungava peninsula. There are Inuit and Cree names for lakes in the area that point to the former presence of

Pacific Harbor Seal, California (Shane Partridge)

Ungava Seal (Doutt 1942)

seals (e.g. *Kasigialuk*, or big place of the harbor seal for Lake Minto). "When the rifle came the coastal harbor seals were heavily exploited, resulting in virtual isolation of the lake seals."

Stories of big fish or cetaceans (grampus) in Ungava lakes may be mythification of seals, according to Harper (1961). For example, Low (1890) includes such in a brief overview of mythical beasts: "Allegorical animals are dreaded and propitiated by these Indians, the greatest among these is the big muskrat who travels under the snow, there is also a big beaver, and a big dog, who does not walk on the ground but upon the trees. In Mistassini is a large trout, so long that he cannot turn round, who causes all the storms on that lake by moving its tail. They have also enormous giants living in the solid rock." Cabot (1920) mentions a Great Grampus "who raised tremendous seas and hauled boats under" in "Ungava Pond," noting also "the term Great Grampus was a loose one, the proper name of this monster of the waters being O-mi-oo-áh-lik (boat wrecker). It lives in Ungava Pond, where the waves are always mountainous. One cannot see across it."

One aspect to Ungava seal habitat will interest those investigating other northern air-breathing freshwater mystery animals. Doutt and Twomey reported (confirmed later by Smith and Horonowitsch (1987)), that at the start of winter, the initial freeze occurs when the lake is high, and the ice is particularly thick and well-anchored near the shore. As the winter progresses, the water levels drop, and there is breakage and refreezing (creating steplike ice sheets) further out from shore where the ice is unsupported. This may not be noticeable on the surface due to snowfall. This freezing pattern creates large air pockets under the ice along the shoreline, which is used by the seals to breathe and even leave the water while being protected from prying eyes above the ice.

Labrador

Given its close proximity to the Ungava seal population, it should be no surprise that similar reports came from lakes in Labrador, right "next door."

Grenfell (1910) related that *Phoca vitulina* is found in many interior freshwater lakes, and "the Indians assert that these freshwater seals never leave the lakes." Strong (1930) noted the presence of seals at Seal Lake, Labrador, based on a paper from 1898 and possibly stories from local Naskapi hunters. Apparently, Doutt also heard stories about seals from these lakes, as there was a brief news release from the summer of 1939, after his 1938 Ungava expedition. It stated:

"Carnegie Museum has extended its sub-Arctic expedition in the wilds of Eastern Labrador a month to investigate reports of a possible new species of seal. The museum was advised by J. Kenneth Doutt, its curator of Mammalogy and co-leader of the expedition, that natives reported curious fresh-water seals, 'larger, redder and of different shape' than those known to science, in the interior of Labrador" (Anon. 1939). Unfortunately, it doesn't appear that Doutt collected anything substantial regarding such seals, as no mention is made in his 1942 monograph.

Interestingly, Strong (1910) also notes the folklore of two mythical otters at Seal Lake and Little Seal Lake: "Two old men of the Davis Inlet band claimed to have seen one of these animals, called *wen-tsúk-ah-més-e-téy-oh*, in Seal Lake (on the Nauscaupee, not the Little Whale River drainage). The body was said to be blackish brown with white lower legs and feet, large ears, and the animal was of great size. It whistled *wheú-u-u*, on a low note, very much like the call of a quail, or so the Indians' rendition of the call sounded. This animal has not been seen for many years, but an old story tells of an Indian who killed the young of this species and was pursued by the mother otter who could swim under land as well as water. According to the story, she killed the man who destroyed her young, but his companion escaped to tell about it. The other mythical otter is called *mis-ín-tsuk*, being about ten feet long and built almost exactly like a seal. It is said not to be dangerous unless attacked. One Davis Inlet man claims to have seen this animal in *atcagon napeesh* or Little Seal Lake (a short distance northeast of Seal Lake). Since harbor seals (*Phoca vitulina*) are fairly common in these lakes, it seems probable that the Indians have combined the characteristics of seal and otter into this mythical conception. They have often killed seals there, but no one has ever killed one of these giant otters." Investigator John Warms (forthcoming) has uncovered sightings in Manitoba regarding large otter-like animals ("water dogs"), suggesting that similar stories might be found in other Canadian provinces.

Nunavut

Robinson (1965) noted in his outdoors column that a reader photographed two seals while fishing at Edehon Lake, which was "at least 110 miles inland from the salt water of Hudson's Bay." The seals were estimated at 200-250 pounds and curious about their plane. (This area, now in Nunavut, was once part of the Northwest Territories.)

Sutton (1965) mentioned in his wildlife column several stories of seals seen over the years in the freshwater lakes along the Thlewiaza River just north of Manitoba. These included Nueltin Lake (in its northern arm), Sealhole Lake, and Edehon Lake. One early report was of a seal shot at Edehon Lake in 1927.

Mansfield (1967) also noted the presence of seals in these lakes along the Thlewiaza River, adding a report by a Caribou Inuit man who saw seals at the north end of Ennadai Lake. Because of these sightings, further study was made, and a brief report was later published (Beck et al 1970) on five seals reported by an aerial survey of the Thlewiaza River region. The report noted that while it was possible that seals lived in the freshwater habitat year round, they were not cut off from Hudson Bay.

Manitoba

Harbor seals are known to travel up the Seal River (near Churchill) for two hundred kilometers or more.

In 1987, a black seal was reported by two individuals in the Winnipeg River. The NRD looked into it, as the sightings were "very detailed" and were from within 40 yards of the animal. They suggested that someone had released it into the area (Anon. 1987).

Dwarf Seals of the Canadian Arctic Islands

Several years ago, I discussed (Arment 2004) a 1932 mention of dwarf seals in the folklore of the Copper Inuit on Victoria Island. Explorer Knud Rasmussen (1932) stated: "They have heard of dwarf seals, which are thought to live in the great lake of the Netsilik: Willistedt Lake. These seals are quite small but just like other seals. They are no bigger than that they can be carried in a game bag, and yet the bull seals smell just the same as real old bull seals." As I noted at the time, similar stories of small seals with mature characters can be found in the United Kingdom.

Since publishing this material, I ran across mention that the Eskimo Museum in Churchill, Manitoba, may in fact have a specimen of a dwarf seal. This was noted in a brief aside in the journal *Museum*, from 1980: "The dwarf seal is extremely rare, and the Eskimo Museum is perhaps the only one to possess a specimen." Details and provenance were not given. I have attempted to contact the current curator of the museum several times, but have not been successful in eliciting a response. Unfortunately, Churchill is off the beaten path, so a simple visit is out of the question. (It can be reached by plane or train, but not by car.) There is local ecotourism (polar bears, whale-watching, and the *Aurora borealis*), though, so perhaps some readers will visit in person and determine what exactly the museum has in its collection.

I also ran across an interesting phenomenon reported in a few seal species in the former Soviet Union (Heptner and Naumov 1996). It was asserted that when some young seals lost their suckling mothers or became ill, if they managed to stay alive, a lack of proper nutrition stunted normal growth. They remained "dwarfs." This was noted anecdotally for both ringed seals and Baikal seals. I haven't seen any other literature on this, so I can't vouch for its legitimacy, but it should be kept in mind as a possible explanation for dwarf seals.

References

Anonymous. 1939. [No Title]. Indiana, Pennsylvania, *Evening Gazette* (August 25): 2.

Anonymous. 1987. Black seal baffles experts. Brandon, Manitoba, *Sun* (December 5): 25.

Arment, Chad. 2004. *Cryptozoology: Science & Speculation*. Landisville, PA: Coachwhip Publications.

Baird, Robin W. 2001. Status of harbour seals, *Phoca vitulina*, in Canada. *Canadian Field-Naturalist* 115: 663-675.

Beck, Brian, Smith, Thomas G., and Mansfield, Arthur W. 1970. Occurrence of the harbour seal *Phoca vitulina* Linnaeus, in the Thlewiaza River, N.W.T. *Canadian Field-Naturalist* 84: 297-300.

Cabot, William B. 1920. *Labrador*. Boston: Small, Maynard & Co.

Chapin, Dave. 2005. In search of the elusive Kasagea: The freshwater seals of central Ungava. *The Tattler* [Keewaydin Temagami] (Fall):3-6.

COSEWIC. 2007. *COSEWIC Assessment and Update Status Report on the Harbour Seal Atlantic and Eastern Arctic subspecies* Phoca vitulina concolor *and Lacs des Loups Marins Subspecies* Phoca vitulina mellonae *in Canada*. Ottawa: Committee on the Status of Endangered Wildlife in Canada.

Dickerson, Scott. 2008. Freshwater seals of Iliamna Lake photographed. Blog. ScottDickerson.com/blog/2008/freshwater-seals-of-iliamna-lake-photographed/

Doutt, J. Kenneth. 1942. A review of the genus *Phoca*. *Annals of the Carnegie Museum* 29(4): 61-125.

Doutt, J. Kenneth. 1954. Observations on mammals along the east coast of Hudson Bay and the interior of Ungava. *Annals of the Carnegie Museum* 33: 235-249.

Dunbar, M. J. 1956. The *Calanus* expeditions in the Canadian arctic, 1947 to 1955. *Arctic* 9(3): 178-190.

Fall, James A., et al. 2006. Subsistence harvests and uses of wild resources in Iliamna, Newhalen, Nondalton, Pedro Bay, and Port Alsworth, Alaska, 2004. *Technical Paper No. 302*. Alaska Department of Fish and Game, Division of Subsistence. Juneau, Alaska: ADFG.

Grenfell, Wilfred T. 1910. *Labrador: The Country and the People*. New York: MacMillan.

Harper, Francis. 1961. Land and fresh-water mammals of the Ungava peninsula. *Miscellaneous Publications* (University of Kansas, Museum of Natural History) 27: 1-178.

Hauser, Donna D. W., et al. 2008. Resident harbor seals (*Phoca vitulina*) in Iliamna Lake, Alaska: Summer diet and partial consumption of adult sockeye salmon (*Oncorhynchus nerka*). *Aquatic Mammals* 34(3): 303-309.

Heptner, V. G., and N. P. Naumov, eds. 1996. *Mammals of the Soviet Union. Vol. II. Part 3. Pinnipeds and Toothed Whales.* Washington, D.C.: Smithsonian & NSF.

Low, A. P. 1890. The Mistassini region. *Ottawa Naturalist* 4:11-28.

Mansfield, Arthur W. 1967. Distribution of the harbor seal, *Phoca vitulina* Linnaeus, in Canadian Arctic Waters. *Journal of Mammalogy* 48(2): 249-257.

Miller, Gerrit S., Jr. 1899. Preliminary list of the mammals of New York. Bulletin of the New York State Museum 6(29): 272-390.

Rainbow Point Lodge. 2011. Fresh water seals. www.akrpl.com/seals.html

Reijnders, Peter, et al. 1993. *Seals, Fur Seals, Sea Lions, and Walrus: Status Survey and Conservation Action Plan.* Gland, Switzerland: IUCN.

Robinson, R. T. 1965. Rod & gun digest. Roblin, Manitoba, *Review.* (September 16): 4.

Smith, Richard J., David M. Lavigne, and William R. Leonard. 1994. Subspecific status of the freshwater harbor seal (*Phoca vitulina mellonae*): A re-assessment. *Marine Mammal Science* 10(1): 105-110.

Smith, T. G., and G. Horonowitsch. 1987. Harbour seals in the Lacs des Loups Marins and eastern Hudson Bay drainage. *Canadian Technical Report of Fisheries and Aquatic Sciences.* No. 1536.

Strong, William Duncan. 1930. Notes on mammals of the Labrador interior. *Journal of Mammalogy* 11(1): 1-10.

Sutton, R. W. 1965. R. W. Sutton's wildlife notebook. Winnipeg, Manitoba, *Free Press* (September 25): 26.

Twomey, Arthur C., and Nigel Herrick. 1942. *Needle to the North: The Story of an Expedition to Ungava and the Belcher Islands*. Boston: Houghton Mifflin Co.

Withrow, David E., and Kymberly M. Yano. 2009. Recent counts of freshwater seals in Alaska's Lake Iliamna. Poster. *Marine Science in Alaska: 2009 Symposium*. Anchorage, AK.

Withrow, David, et al. 2011. Freshwater harbor seals of Lake Iliamna, Alaska: Do they pup and over-winter in the lake? Poster. *2011 Alaska Marine Science Symposium*. Anchorage, AK.

A Bipedal Reptile in Nevada

Chad Arment

Accounts of bipedal reptiles ("dinosaur"-like or otherwise) have been sporadically reported in North America for many years. Most sightings come from western states, and many have been collected and documented over the last few years by Nick Sucik and myself (Sucik 2006; Arment 2008). Nick has had opportunity to travel to several regions with alleged sightings, and has interacted with local news sources in an attempt to solicit further witnesses.

In December 2009, Nick was contacted by a witness (JF) who had read an online newspaper article about his research. JF's initial email included the following (lightly edited):

> "My daughter and I were held in amazement on June 28th, 2009. It was 5:50pm. (I know because we needed to get gas with cash and the station closes at 6:00. The attendant was on his way out the door, cash box in hand when we pulled in!) The day was warm and sunny, with a slight breeze from the southwest. What we saw was not a bird, but a bipedal, reptilian animal with a few bird type features. It was very small, the size of a chukar, with strong looking back legs, frilly plume-looking feather things on places, two fore legs with small claws, little spaced teeth, a tail about half as long as it was that it carried just above ground height. But, the one thing I remember thinking remarkable was it did a full basketball player pivot when it turned to run. It's no emu chick or monitor lizard. It was 20 feet away in full

sun. Beautiful. ... We know it sounds crazy and I guess I wish it had happened to someone else! But any info would be appreciated."

At the time, circumstances did not allow Nick to immediately respond, so he asked me to get in touch with her. I spoke with JF on the phone for over an hour, and she impressed me as knowledgeable, curious, and excited to talk to someone about the incident without being ridiculed.

JF is a California schoolteacher, and she and her husband had bought land up in the Nevada mountains to build a cabin. They had been taking trips there for the last year or so, camping on the property. This trip, she and her daughter were alone, coming back in the middle of the day. They were heading south just outside the town of Gerlach, Nevada. Gerlach is in east-central Washoe County. Gerlach is a very small town in a sparsely inhabited part of the state. The only time they get many visitors is when the annual week-long Burning Man festival is held nearby.

Her daughter had just rolled down the windows to air the car out due to some dog hair coming through the vents, and JF slowed down as they reached a large curve in the road. She saw a large rabbit lying dead in the middle of the opposite lane. (She didn't see any signs that the animal had been hit, but described it as looking like it had just sat down, rolled over, and died.) She was only going about 15 mph at this time, when she saw this rabbit and thought it was strange that it hadn't been picked off by scavengers. Then she saw a reptilian head pop up from the opposite side of the rabbit—it had its eyes closed and its mouth open. She was barely going at a crawl by this time, and was about 20 feet away with the sun directly behind her, so she had a very good look at the animal.

It opened its eyes, saw the truck, threw its forelegs up into the air, pivoted and dashed off into the sagebrush. Her daughter, sitting next to her, looked out the back and watched it disappear as they drove past, saying, "That's not a bird!" JF had slowed down

Witness' sketch, used with permission.

Witness' sketch, used with permission.

when she saw the animal, thinking “it was cute at first,” but when it ran off, “it wasn’t cute anymore,” so she didn’t stop.

She described it as approximate 15 to 18 inches tall when standing upright, with a similar length. (She said that it would have had trouble dragging off the rabbit.) The tail was kangaroo-like, more like a rudder than a long whip-like tail. The tail stood straight out, balancing, as it ran. The head was reptilian, but not narrow in snout (so, not like a caiman), and lacking a beak. The head was about the size of a chihuahua’s head. The neck wasn’t very long. Its mouth being open, she noted 4-6 small triangular teeth on each side, sort of separated between. The eye was yellow, and she noticed the pupil was vertical (up and down) in orientation. She didn’t notice the hands herself, but her daughter said they had small fingers and claws. The arms were long enough to reach the head, and had some musculature, but didn’t look like they’d be used quadrupedally. The hind legs were stout and muscular, but she didn’t see the feet. She didn’t notice any scales, and the skin seemed smooth.

The head and body seemed to be covered with fluffy spiky tassle-like hairs or feathers. They were sort of stiff but flexible (but not like down). The wind was blowing and parted the hairs on the head “like an Afro.” The coloration of the hairs seemed to match that of the body: the head was burgundy-reddish in color, while the body was a light tan to pearlescent green. She thought it was really pretty. As soon as it dashed off into the sagebrush, the colors blended in like perfect camouflage, and the spiky hairs helped it disappear into the surrounding sparse brush.

She said it didn’t act like a normal lizard, as it seemed more aware and wary of them than lizards would be.

JF is a schoolteacher (and her parents were biology teachers), she has neighbors who keep emus, and she keeps chickens, and she is familiar with a variety of birds and reptiles. She said this clearly wasn’t anything she recognized. She said the locals in the town never go out into the desert except to hunt. She noted it is pretty cold there in winter, and very hot in the summer.

When evaluating an eyewitness account of a strange animal, the investigator must keep in mind several possibilities:

1) The witness was truthful and accurately presented details.
2) The witness was truthful, but may have reported inaccurate or mistaken details.
3) The witness was not truthful, and deliberately provided incorrect details.

No investigator wants to be hoaxed, so we look at this possibility first. To counter this, we have several lines of evidence. First, the witness isn't anonymous. She provided Nick and I with full contact information. Second, she wasn't seeking out media attention. She had looked online to see if there were similar reports, and posted her story to at least one page, but she said some of the UFO/Creature sites she looked at were too weird for her, so she didn't do more along those lines. Third, as I spoke with her, she first told her story almost nonstop for forty minutes, barely stopping to recall anything, before I began asking questions. This is a good indication that the sighting actually took place and wasn't made up as she went along. Fourth, she didn't jump to the most controversial possible identity for this animal—dinosaurian. She thought it might just be some kind of strange lizard, though she was familiar with common species like collared lizards, and said it wasn't similar to them. I am certain that JF is not attempting a hoax. Speaking with her again, almost a year after my initial contact, she is still very interested in anything that can be learned about this animal.

Regarding the possibility of inaccurate details, relying on human memory is always tricky. In this case, there were two witnesses, and both immediately recognized the unusual nature of the animal. They made a sketch of the animal after arriving home, in order to keep from forgetting details. They saw the animal in good light from not too far a distance, but of course, we have to weigh surprise as a possible influence on both memory and interpretation of details when the animal was first seen. (In this case, at least, there was no fear of the creature. JF's first thought was that it was cute, not scary.)

The potential for misidentification can never be ruled out, though there doesn't appear to be a reasonable candidate that immediately stands out. I don't think a known native lizard is likely, and even exotics like varanids are not a good fit. A bird would be more likely than a lizard. I would like to see more than an *ad hoc* proposal stretching to fit a "foregone" conclusion, however. For now, I consider this a legitimate cryptozoological sighting.

Of course, even if we assume that a witness is accurate in most details, this doesn't identify or scientifically describe a creature. It just gives us a few more clues that could be worth following up in order to find enough evidence to physically confirm it as a new (and perhaps surprising) species. The points of particular interest in this case include aspects of diet (scavenging), appearance (ability to camouflage itself with both color and morphological characters), and posture (the rudder-like tail). And if it does represent something new and distinctive, there is no certain way to determine whether it is a juvenile form or adult form, though of course speculation could be made.

Rather than speculation, what is needed is serious field investigation in this and nearby regions with similar reports, but such would no doubt be expensive and time-consuming. It would probably be easier to solicit help from local individuals, but even that can be problematic, particularly with something as controversial and far from mainstream (even within cryptozoology) as a "dinosaur-like" animal here in North America.

References

Arment, Chad. 2008. An account of a North American bipedal reptile from Wyoming. *BioFortean Review* No. 19. http://www.strangeark.com/bfr/articles/wyoming-bipedal-lizard.html

Sucik, Nick. 2006. "Dinosaur" sightings in the United States. in Chad Arment, ed. *Cryptozoology and the Investigation of Lesser-Known Mystery Animals*. Landisville, PA: Coachwhip Publications.

Coachwhip Publications

CoachwhipBooks.com

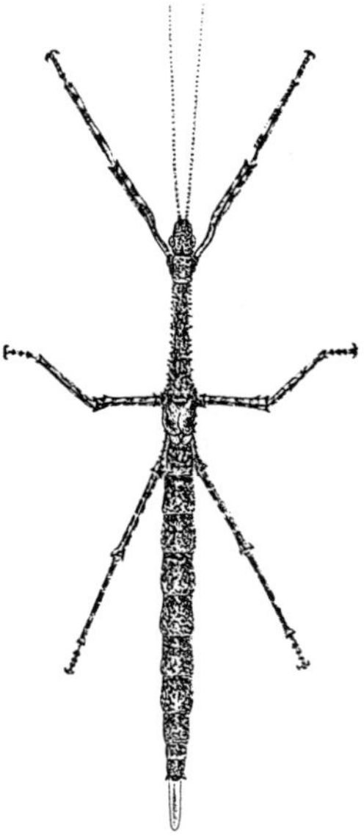

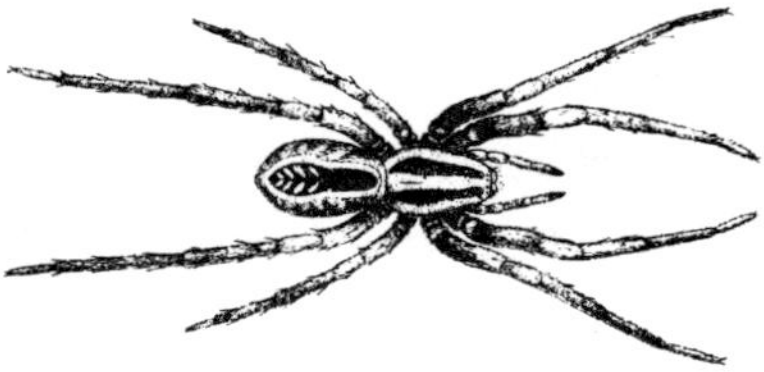

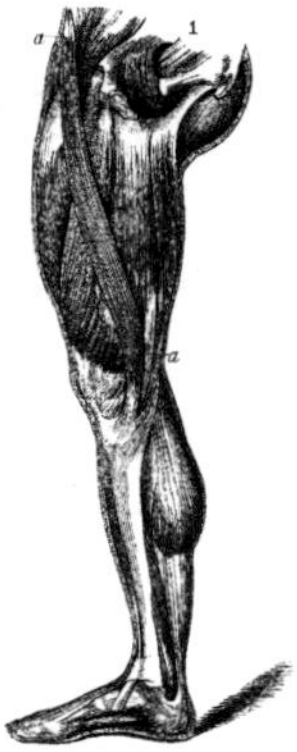

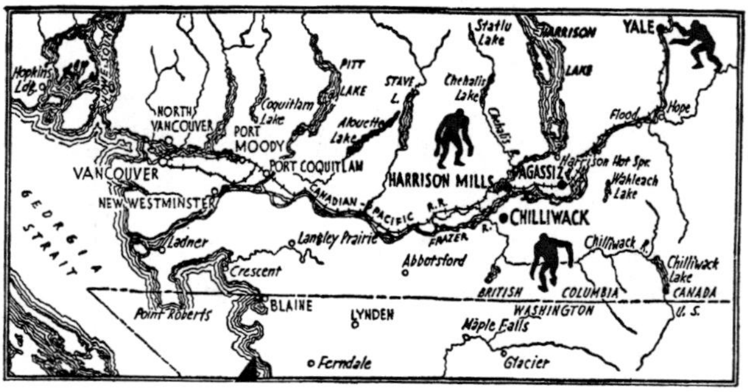

Coachwhip Publications

CoachwhipBooks.com

BioFortean Reprints

Coachwhip Publications

CoachwhipBooks.com

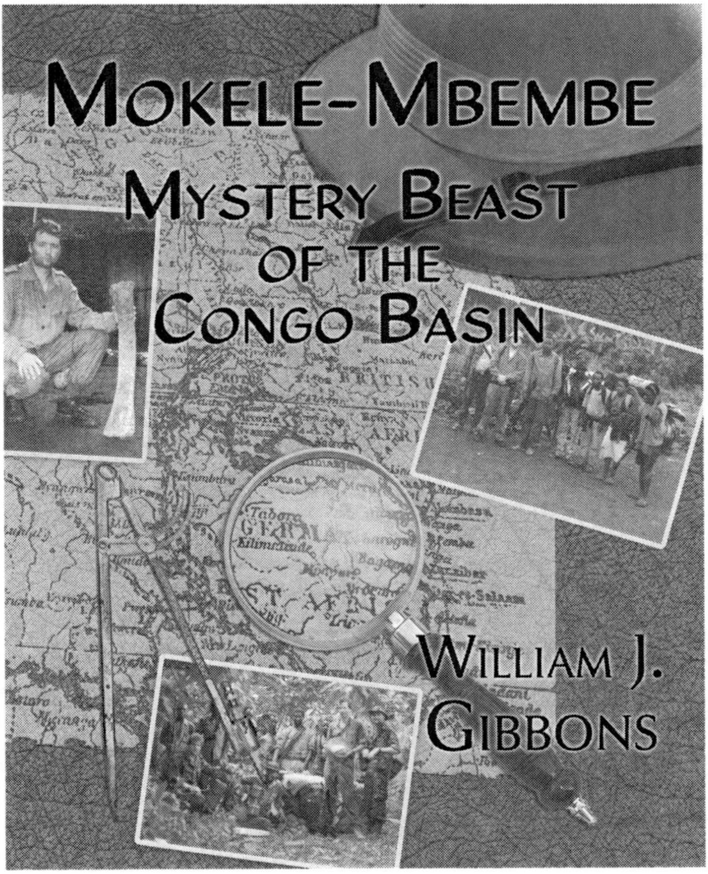

In search of a dinosaur-like creature in the depths of the Congo river basin . . .

CPSIA information can be obtained at www.ICGtesting.com
Printed in the USA
LVOW130854170812

294736LV00005B/15/P

9 781616 460877